PEPTIDE MADE PRACTICAL

Transform Your Health with Peptides Anti-Aging, Injury-Free Muscle Support, and Mental Clarity Science-Proven Protocols for Lasting Wellness

Cassian King

© 2024 Peptide Made Practical All rights reserved. This document is intended exclusively for informational use in connection with this book. The book is provided "as is" without any warranties, whether express or implied. Unauthorized reproduction, distribution, or transmission of this book, in whole or in part, is strictly forbidden. All trademarks and brand names mentioned herein are the property of their respective owners. The publisher disclaims any responsibility for damages that may arise from the use or misuse of the information contained in this book.

Table of Contents

Introduction 8

    Why Choose Peptides for Health 8

    History and Discovery of Peptides 9

    Book Goals and Responsible Use Guidelines 10

Chapter 1: Introduction to Peptides 12

    Definition and Function of Peptides 12

    Differences Between Peptides, Proteins, Amino Acids 13

    Mechanisms of Action 14

Chapter 2: The Science of Cellular Aging 17

    Cellular Senescence and Rejuvenation 17

    Cellular Renewal and Anti-Aging 18

    Peptides and Longevity Studies 20

Chapter 3: Peptides for Brain Health 23

    Peptides for Cognitive Enhancement 26

        Preventing Cognitive Decline 27

    Regulating Sleep and Reducing Stress 29

        Peptides for Sleep and Anxiety 31

Chapter 4: Peptides for Muscle Recovery 33

    Tissue Repair and Recovery Peptides 33

        Peptides for Tissue Regeneration 35

    Safe Muscle Growth with Peptides 36

        Anabolic Peptides and Safe Use Cycles 38

        Combination Strategies for Muscle Growth 40

Chapter 5: Peptides for Weight Loss 42

Lipolytic Peptides for Fat Loss *42*

    Lipolytic Peptides and Usage Methods 43

Appetite Regulation Peptides *45*

Peptides for Appetite Regulation *46*

Audio Version 49

Chapter 6: Peptides for Skin Health 50

Improving Skin Elasticity and Texture *50*

    Peptides for Reducing Wrinkles 51

Skin Renewal and Care *53*

Chapter 7: Safe Peptide Use Guidelines 56

Dosages and Administration Guide *56*

    Recommended Dosages and Administration 57

    Administration Methods and Effectiveness 58

Cycles and Treatment Durations *60*

    Planning Usage Cycles 62

    Cycle Strategies for Benefits and Risks 64

Chapter 8: Combined Protocols and Synergies 67

Synergistic Peptides *67*

    Effective Health Combinations 69

Avoiding Negative Interactions *70*

    Choosing Safe and Compatible Peptides 72

Chapter 9: Monitoring and Optimizing Results 75

Measuring Protocol Effectiveness *75*

    Evaluating Peptide Impact Techniques 76

Personalized Protocol Adjustments *78*

    Modifying Protocols for Personal Results 80

Chapter 10: Potential Risks and Side Effects 82

Common Side Effects *82*

Signs and Symptoms to Monitor 83

Safety Warnings and Precautions 85

    Importance of Consulting a Doctor 87

Contraindications in Peptide Use 88

    Health Conditions Requiring Attention 90

Chapter 11: Peptide Legality and Regulation 92

Peptide Regulations in the USA 92

    Buying Peptides Safely and Legally 94

Buying Peptides: Quality and Safety Tips 95

    Safety and Reliable Peptide Sources 97

Chapter 12: Peptides in Nutrition and Lifestyle 100

Diet to Support Peptide Benefits 100

    Essential Nutrients for Peptide Benefits 101

Exercise and Rest Regime 103

    Fitness, Sleep, and Peptide Effectiveness 105

Complete Wellness Strategies 107

    Integrating Peptides into Health Routines 108

Chapter 13: Testimonials and Case Studies 111

Real Experiences with Peptides 111

    Success Stories in Health Improvement 112

Clinical Case Studies on Peptide Efficacy 113

    Peptide Health Improvement Cases 115

Conclusion 118

Final Thoughts on Peptide Use 118

Holistic Long-Term Approach 119

Conscious Health Journey 121

Appendices 122

Glossary of Technical Terms 122

Additional Resources ... 123

Bibliography ... 124

Analytical Index ... 125

YOUR 3 BONUS IS WAITING FOR YOU!

I'm thrilled to gift you **three additional guide**:

- **Nutrition & Lifestyle for Peptide Results** – A practical guide on the best nutritional and lifestyle choices to maximize your results with peptides.
- **Top 10 Anti-Aging Hacks** – Discover simple, effective hacks for reducing the signs of aging and supporting your body's natural regenerative processes.
- **The Peptide Travel Guide** – Everything you need to know to safely and effectively manage your peptide routine while traveling.

👇 SCAN HERE TO DOWNLOAD IT

Introduction

Why Choose Peptides for Health

Peptides, short chains of amino acids linked by peptide bonds, have emerged as a cornerstone in the pursuit of health and wellness, particularly for those keen on anti-aging, muscle recovery, and mental clarity. Unlike proteins, peptides are smaller and more easily absorbed by the body, allowing them to act quickly and efficiently in various physiological processes. Their role is pivotal in signaling cells to perform specific functions such as healing, rejuvenation, and even stimulating the body's natural production of growth hormone, which is crucial for maintaining muscle mass and a healthy metabolism.

For individuals passionate about maintaining an active lifestyle and those concerned with the natural decline in energy levels and physical capabilities that come with aging, peptides offer a scientifically backed solution. They support the body's recovery process after exercise, reduce inflammation, and contribute to skin elasticity and overall vitality. This makes peptides an invaluable tool for anyone looking to enhance their physical performance, accelerate recovery times, and achieve a more youthful appearance without resorting to invasive procedures.

Moreover, the mental clarity benefits associated with specific compounds cannot be overstated. In a world where cognitive function is paramount, these substances provide a means to enhance focus, memory, and cognitive resilience, thereby supporting both mental and physical aspects of wellness.

Given the cautious nature of health-conscious individuals regarding expenditure on wellness products, the appeal of peptides also lies in their evidence-based efficacy. Backed by scientific research, peptides offer a tangible return on investment in terms of health and wellness benefits, making them a prudent choice for those dedicated to optimizing their health.

Incorporating peptides into personal health regimes is not merely about addressing current concerns but is a proactive approach to long-term wellness. By leveraging the targeted actions of these bioactive compounds, individuals can support their body's natural processes, from muscle recovery to cognitive function, ensuring a holistic approach to health that aligns with their active and wellness-focused lifestyle.

History and Discovery of Peptides

The exploration and understanding of peptides began in the early 20th century, marking a pivotal moment in the field of biochemistry and medicine. Initially, peptides were identified as simple chains of amino acids, the basic building blocks of proteins. This discovery was crucial, as it shed light on how these molecules play a significant role in various biological processes. The term "peptide" itself stems from the Greek word "peptós," meaning "digested," pointing to the early recognition of peptides' involvement in digestion. Over time, scientists began to unravel the complex roles of peptides beyond digestion, identifying them as key players in signaling, immune responses, and cellular functions.

One landmark moment in research was the discovery of insulin in the 1920s, a hormone that revolutionized the treatment of diabetes. This discovery not only highlighted the therapeutic potential of these biomolecules but also opened the door to the concept of therapeutic applications involving them. Following this, there was a surge in research, leading to the identification of numerous small proteins with critical physiological functions, including hormones, neuropeptides, and antimicrobial compounds, among others.

The technological advancements in synthesis during the latter half of the 20th century further propelled the field forward. Solid-phase synthesis, developed in the 1960s, allowed for the more efficient production of these biomolecules, making it possible to create specific compounds for research and therapeutic use. This method became a cornerstone in the development of drug-based therapies.

In recent years, the interest in these small proteins has surged, driven by their potential in addressing a wide range of health issues from metabolic disorders to aging. The discovery of bioactive compounds with anti-aging, muscle recovery, and cognitive function benefits has led to a new era of wellness protocols. These advancements underscore the significant potential in not only treating but also preventing various health conditions, marking them as a key component of future therapeutic and wellness strategies.

Book Goals and Responsible Use Guidelines

The primary objective of this book is to demystify the complex world of peptides and present practical, science-backed protocols that can be seamlessly integrated into daily life for improved health, wellness, and longevity. With a focus on anti-aging, muscle recovery, and mental clarity, the guidance offered is aimed at empowering readers, especially health-conscious women aged 30 to 50, with the knowledge and tools necessary to make informed decisions about peptide use. Given the diversity in educational backgrounds and the varying degrees of familiarity with biochemical concepts among our readers, every effort has been made to ensure that the information is accessible, clear, and actionable.

Responsible use of peptides is paramount. It involves understanding the correct dosages, recognizing the importance of purity in peptide selection, and acknowledging the need for professional medical advice before beginning any new supplement regimen. Peptides, while offering numerous health benefits, are potent biological molecules, and their use should be approached with the same caution and diligence as any other health intervention. This includes awareness of potential side effects, the importance of sourcing from reputable suppliers, and the necessity of adhering to legal guidelines regarding their purchase and use.

To ensure safety and efficacy, it is crucial to follow **recommended dosages** and **administration methods**, which have been carefully outlined in subsequent chapters. Equally important is the understanding of **cycles and treatment durations**, to maximize the benefits of peptides while minimizing any risks. Personalization of peptide protocols, based on individual health status, goals, and responses, is encouraged, with continuous monitoring and adjustments as needed.

This book also emphasizes the significance of a holistic approach to health and wellness, where these bioactive compounds are integrated into a broader lifestyle strategy encompassing a balanced diet, regular exercise, and adequate rest. The synergy between peptides and these foundational pillars of health can amplify the benefits and contribute to lasting wellness.

By adhering to these guidelines and adopting a responsible, informed approach to peptide use, readers can look forward to achieving their health and wellness objectives safely and effectively.

Chapter 1: Introduction to Peptides

Definition and Function of Peptides

Peptides are essentially short chains of amino acids, the building blocks of proteins, but they are distinguished by their shorter length and specific functions within the human body. Unlike proteins, which consist of long chains of amino acids and serve as fundamental components of tissues and organs, peptides typically contain between 2 to 50 amino acids. This smaller size allows peptides to be absorbed more easily by the body, enabling them to enter the bloodstream and reach their target sites more efficiently. In the human body, these small chains of amino acids play a crucial role in a wide range of biological processes, including hormone production, cell signaling, and enzymatic reactions. They can act as neurotransmitters, hormones, or even as key regulators of various physiological processes such as inflammation, immune response, and the repair and regeneration of tissues.

One of the primary functions of peptides in the body is to serve as signaling molecules. They bind to receptors on the surface of cells, triggering specific responses that influence the behavior of those cells. For instance, some peptides stimulate the production of growth hormone, which plays a vital role in muscle growth, metabolism, and tissue repair. This signaling function is particularly relevant in the context of health and wellness, where these biomolecules are used to promote healing, reduce inflammation, and support the body's natural recovery processes.

Additionally, these small chains of amino acids contribute to the body's defense mechanisms. Antimicrobial compounds, for example, possess the ability to fight off infections by destroying harmful bacteria and viruses. This protective role underscores the potential of these molecules not only in maintaining health but also in preventing disease.

In the realm of anti-aging and wellness, certain compounds have been identified for their ability to enhance skin elasticity, reduce the appearance of wrinkles, and support overall skin health. These cosmetic agents work by stimulating collagen production, a key protein that maintains skin structure and firmness. Moreover, substances involved in cognitive health can improve mental functions such as memory, focus, and cognitive agility, thereby addressing concerns related to aging and mental clarity.

The metabolic functions of these biomolecules also play a significant role in weight management and energy balance. They can influence appetite, metabolism, and fat burning, making them valuable tools in weight loss and body composition efforts. By modulating these physiological processes, peptides can help individuals achieve their health and fitness goals, contributing to a balanced and active lifestyle.

Understanding the diverse functions of these biological molecules in the human body reveals their potential as powerful allies in the pursuit of health and wellness. From supporting muscle recovery and promoting skin health to enhancing cognitive function and aiding in weight management, the targeted actions of these compounds offer a promising approach to addressing a variety of health concerns. As research continues to uncover new molecules and their functions, the possibilities for health optimization and disease prevention appear increasingly promising, making them a key focus for those dedicated to a proactive and informed approach to personal wellness.

Differences Between Peptides, Proteins, Amino Acids

To grasp the fundamental distinctions between peptides, proteins, and amino acids, it's essential to delve into their structure and function within the human body. At the core, amino acids are the building blocks of both peptides and proteins, serving as the smallest units in this biochemical hierarchy. There are twenty standard amino acids, each with a unique side chain that determines its properties and role in protein synthesis. When amino acids link together through peptide bonds—a type of covalent bond formed between the amino group of one amino acid and the carboxyl group of another—they form peptides. This linkage is a dehydration synthesis reaction, where the removal of a water molecule facilitates the bond.

Short chains of amino acids, typically fewer than 50, are classified as such, whereas longer chains are considered proteins. This distinction is not merely quantitative but also reflects differences in complexity and functionality. Due to their smaller size, these molecules tend to be more involved in signaling and regulatory functions within the body. They can easily penetrate tissues and cells, allowing them to act swiftly in various physiological processes such as hormone secretion, immune responses, and cell signaling pathways. For example, insulin, a well-known hormone made up of amino acids, plays a critical role in regulating glucose metabolism by facilitating the uptake of glucose into cells.

Proteins, on the other hand, are larger molecules that often fold into complex three-dimensional structures, which are crucial for their function. These structures enable proteins to take on a variety of roles in the body, including structural support, as seen in collagen; movement, as with actin and myosin in muscle tissue; catalyzing biochemical reactions as enzymes; and transporting molecules, exemplified by hemoglobin's role in transporting oxygen through the blood. The functionality of proteins is heavily dependent on their structure, which in turn is determined by the sequence and arrangement of amino acids within the molecule. Any alteration in the sequence can significantly impact a protein's function, as seen in many genetic disorders caused by mutations in DNA that lead to incorrect amino acid sequences in proteins.

Understanding the differences between peptides, proteins, and amino acids is crucial for comprehending their diverse roles in the body and their potential therapeutic applications. While amino acids serve as the essential building blocks, peptides and proteins, through their unique structures and sizes, execute a wide array of functions that are vital for maintaining health and homeostasis. This knowledge forms the foundation for exploring how peptides can be harnessed for health and wellness, particularly in the realms of anti-aging, muscle recovery, and cognitive function. By leveraging the targeted actions of peptides, individuals can support their body's natural processes, ensuring a holistic approach to health that aligns with their active and wellness-focused lifestyle.

Mechanisms of Action

Peptides exert their influence on the body through various mechanisms of action, each tailored to specific cellular functions and therapeutic targets. At the core of their functionality, peptides interact with cells by binding to specific receptors on the cell surface, initiating a cascade of intracellular events that can alter cellular behavior, promote healing, and enhance physiological functions. This receptor-ligand interaction is fundamental to understanding how peptides achieve their effects, ranging from anti-aging to muscle recovery and mental clarity.

One of the primary mechanisms by which these compounds affect the body is through the modulation of signaling pathways. For instance, certain substances can mimic the action of growth hormone-releasing hormone (GHRH), leading to an increase in the body's production of growth hormone. This elevation in growth hormone levels can have profound effects on tissue repair, muscle growth, and metabolism, contributing to improved muscle recovery and a reduction in body fat. The specificity of compound-receptor interactions ensures that these substances can exert these effects with a high degree of precision, minimizing the risk of off-target effects and unwanted side effects.

Another significant mechanism is the regulation of cellular differentiation and proliferation. Peptides can influence the fate of stem cells, encouraging their differentiation into specific cell types needed for tissue repair and regeneration. This capability is particularly valuable in the context of injury recovery and anti-aging, where the efficient replacement of damaged or aged cells is crucial for maintaining tissue function and integrity.

These biomolecules also play a critical role in modulating the immune response. By affecting the activity of various immune cells, they can help reduce inflammation, a key factor in many chronic diseases and age-related conditions. Anti-inflammatory compounds can support the healing process by mitigating the detrimental effects of excessive inflammation, thereby accelerating recovery from injuries and reducing the risk of inflammatory diseases.

In the realm of brain health, certain peptides can cross the blood-brain barrier, a unique capability that allows them to exert neuroprotective effects. These peptides can enhance cognitive function, protect neurons from damage, and support the overall health of the nervous system. By influencing neurotransmitter release and synaptic plasticity, peptides can improve mental clarity, memory, and mood, offering potential benefits for individuals experiencing cognitive decline or seeking to optimize their cognitive performance.

Understanding these mechanisms of action provides a foundation for the responsible and effective use of these bioactive compounds in health and wellness. By targeting specific cellular pathways and functions, these substances offer a versatile and targeted approach to improving health, with the potential for personalized therapies tailored to the unique needs of each individual. Through ongoing research and clinical studies, the full spectrum of their therapeutic potential continues to be explored, promising new possibilities for disease treatment, health optimization, and the enhancement of human performance. This exploration not only highlights the importance of these compounds in addressing various health concerns but also emphasizes the need for a nuanced understanding of their interactions within the body, paving the way for innovative applications in both preventive and restorative health strategies. As we delve deeper into the science behind these compounds, we uncover their roles in modulating biological processes, which can lead to breakthroughs in how we approach health and wellness in a holistic manner. The future of health optimization lies in our ability to harness these mechanisms effectively, ensuring that individuals can achieve their health goals through informed and tailored interventions.

Chapter 2: The Science of Cellular Aging

Cellular Senescence and Rejuvenation

Cellular senescence is a state in which cells cease to divide but do not die, remaining metabolically active. This process is a double-edged sword in the context of human health and aging. On one hand, senescence plays a crucial role in preventing the proliferation of damaged cells, thus acting as a natural barrier against cancer. On the other hand, the accumulation of senescent cells contributes to aging and age-related diseases by secreting pro-inflammatory factors that lead to tissue dysfunction. The concept of cellular rejuvenation focuses on strategies to remove or modify senescent cells, thereby potentially reversing age-related decline and enhancing tissue repair and regeneration.

Recent scientific advancements have highlighted peptides as promising agents in the fight against cellular senescence and the quest for rejuvenation. Peptides, due to their ability to modulate cellular processes, offer a targeted approach to influence senescent cells. Certain peptides can selectively induce apoptosis in senescent cells, a process known as senolytic activity. This selective removal of senescent cells can mitigate their detrimental effects on tissue health and function, potentially delaying the onset of age-related diseases and extending healthspan.

Moreover, these bioactive compounds can also enhance the body's repair mechanisms. By stimulating pathways involved in cell proliferation and differentiation, they support the replacement of senescent cells with healthy ones, promoting tissue regeneration. This is particularly relevant in the context of skin health, muscle recovery, and overall tissue integrity, areas of significant concern for the health-conscious individual.

In addition to their direct effects on cellular senescence, these compounds can modulate the immune system, reducing chronic inflammation, which is both a cause and effect of cellular senescence. By attenuating the inflammatory environment, they can further support tissue health and longevity.

The practical application of these substances for cellular rejuvenation involves careful consideration of dosage, administration routes, and treatment schedules. Research suggests that the benefits of these agents are dose-dependent, with specific ones requiring precise administration protocols to achieve the desired senolytic or rejuvenative effects. Furthermore, the combination of these agents with other interventions, such as lifestyle modifications and nutritional support, may enhance their efficacy in promoting cellular rejuvenation.

It is essential for individuals interested in leveraging these compounds for anti-aging and rejuvenation to consult with healthcare professionals experienced in this therapy. A personalized approach, taking into account the individual's health status, goals, and potential contraindications, is crucial for maximizing the benefits of these interventions while minimizing risks.

The exploration of these substances as agents of cellular senescence and rejuvenation represents a frontier in anti-aging medicine. With ongoing research and clinical trials, the potential of these compounds to support health and wellness by targeting the cellular underpinnings of aging continues to expand, offering hope for effective, science-based strategies to enhance longevity and quality of life.

Cellular Renewal and Anti-Aging

Peptides, as crucial components in the cellular renewal process, offer a promising avenue for anti-aging interventions. Their role in promoting cellular health and longevity is deeply rooted in their ability to modulate various biological pathways, essentially instructing cells to behave in a more youthful manner. This capability not only highlights the potential of peptides in slowing down the aging process but also underscores their significance in enhancing the body's natural regenerative processes. By stimulating collagen production, peptides directly contribute to the improvement of skin texture and elasticity, addressing one of the most visible signs of aging. Collagen, a primary structural protein in the skin, tends to diminish as we age, leading to wrinkles and loss of firmness.

Peptides signal the skin cells to synthesize more collagen, thereby promoting a more youthful appearance.

Furthermore, these bioactive compounds play a pivotal role in muscle repair and growth. This is particularly beneficial for individuals engaged in regular physical activity, as the ability to recover from muscle injuries or strains is crucial. Substances such as growth hormone-releasing agents (GHRAs) and selective androgen receptor modulators (SARMs) have shown promise in enhancing muscle mass and strength by mimicking the effects of natural growth hormone, without the associated side effects of synthetic growth hormone administration. This not only aids in maintaining an active lifestyle but also contributes to a healthier metabolism, which is often affected by aging.

The cognitive benefits of peptides cannot be overstated. Certain peptides have the ability to cross the blood-brain barrier, offering neuroprotective effects that can enhance cognitive function, memory, and mental clarity. This aspect of peptide therapy is of significant interest, as cognitive decline is a common concern among aging populations. By supporting neuron health and facilitating better communication between brain cells, peptides offer a potential strategy for maintaining cognitive abilities well into later life.

On a cellular level, these bioactive compounds influence longevity through their impact on telomere length. Telomeres, which are protective caps at the end of chromosomes, shorten with each cell division, and their length is directly associated with cellular aging. Some of these compounds have been shown to slow the rate of telomere shortening, thereby extending the lifespan of cells. This not only has implications for longevity but also for the overall health and functionality of organs and tissues throughout the body.

The anti-inflammatory properties of these substances also contribute to their anti-aging effects. Chronic inflammation is a known driver of aging and age-related diseases. By modulating the immune response and reducing inflammation, these compounds can help mitigate the risk of conditions such as heart disease, diabetes, and neurodegenerative diseases, all of which are more prevalent in older populations.

In considering the use of peptides for cellular renewal and anti-aging, it is essential to approach their application with a personalized strategy. Factors such as individual health status, specific aging concerns, and desired outcomes should guide the selection of peptides and the design of therapy protocols. This tailored approach ensures that the benefits of peptides are maximized, addressing the unique needs of each individual while minimizing potential risks.

The integration of peptides into a comprehensive wellness regimen, including a balanced diet, regular exercise, and adequate rest, can amplify their effects. Such a holistic strategy not only supports the efficacy of peptide therapy but also promotes overall health and well-being, laying a solid foundation for a healthier, more vibrant aging process.

Peptides and Longevity Studies

The burgeoning field of longevity research has increasingly spotlighted these biomolecules as pivotal agents in extending lifespan and enhancing healthspan. Given their intrinsic role in cellular communication, repair, and regeneration, they have emerged as key players in modulating aging processes. Scientific studies and theoretical frameworks have provided compelling evidence on how these compounds can influence longevity, primarily through mechanisms such as telomere extension, autophagy promotion, and senescence regulation. These pathways, critical to the aging narrative, offer a glimpse into how such substances may pave the way for not just a longer life, but one marked by vigor and diminished age-associated decline.

Telomere Extension: Telomeres, the protective caps at the ends of chromosomes, naturally shorten with each cellular division, a process intrinsically linked to cellular aging. The enzyme telomerase can extend these telomeres, potentially slowing the aging process. Certain peptides have been identified as telomerase activators, suggesting their role in promoting longer telomeres and, by extension, influencing longevity. Studies in cell cultures and animal models have shown that peptides stimulating telomerase activity can lead to extended cellular lifespan, offering a tantalizing hint at their potential for human aging.

Autophagy Promotion: Autophagy, the body's mechanism for cleaning out damaged cells, is another critical process influenced by peptides. By removing these cells, autophagy prevents bad proteins from accumulating and causing diseases. Peptides have been shown to enhance autophagy, thus potentially reducing the risk of age-related diseases and contributing to a longer healthspan. This process is vital for maintaining cellular health and function, with implications for longevity by preserving the body's ability to renew itself.

Senescence Regulation: Cellular senescence is a state where cells cease to divide but do not die, contributing to aging and age-related diseases. Peptides that can selectively induce apoptosis in senescent cells or inhibit their senescence-inducing pathways offer a strategy for mitigating the negative impacts of senescent cells on tissue function and longevity. By clearing or preventing the accumulation of these cells, peptides can potentially delay aging and extend healthspan, offering a promising avenue for anti-aging interventions.

Mitochondrial Function: The mitochondria, known as the powerhouse of the cell, play a crucial role in energy production and are involved in aging and longevity. Peptides that support mitochondrial function can help maintain energy levels, reduce oxidative stress, and prevent mitochondrial dysfunction, all of which are associated with aging. Enhancing mitochondrial efficiency through peptide therapy could thus contribute to improved vitality and extended lifespan.

Inflammation Reduction: Chronic inflammation is a hallmark of aging and is implicated in the development of many age-related diseases. Compounds with anti-inflammatory properties can play a significant role in modulating the immune response, reducing inflammation, and thereby potentially mitigating the impact of aging on the body. By targeting inflammation, these substances offer a pathway to not just longer life, but one less burdened by the diseases of aging.

The exploration of these compounds in the context of longevity is supported by a growing body of research, yet it is important to approach this promising field with a balanced perspective. While preclinical studies have shown significant potential, the translation of these findings into human health and longevity remains a complex challenge. Ongoing clinical trials and longitudinal studies are crucial to fully understand the role of these substances in human aging and to establish evidence-based protocols for their use in longevity medicine.

Furthermore, the integration of this therapy into a holistic approach to health, encompassing diet, exercise, and lifestyle modifications, is likely to amplify the benefits in promoting longevity. As research continues to unveil the mechanisms through which these compounds influence aging, there is cautious optimism that such interventions will become a cornerstone of age-related healthcare, offering individuals pathways to not only extend their lifespan but to enhance their healthspan, embodying a future where aging is not synonymous with decline.

Chapter 3: Peptides for Brain Health

Peptides, with their vast array of functions within the human body, have garnered significant attention for their potential to enhance brain health, cognitive function, and mental clarity. These short chains of amino acids are capable of crossing the blood-brain barrier, a unique characteristic that enables them to exert direct effects on the central nervous system. This capability is particularly important for addressing cognitive decline, enhancing memory, and supporting overall brain health. Among the peptides of interest for these purposes, nootropic peptides stand out for their ability to improve cognitive functions. These peptides can influence neurotransmitter levels, enhance neuron communication, and support neuroprotection, thereby offering a potential therapeutic strategy for cognitive enhancement and the mitigation of age-related cognitive decline.

Another significant area of interest is the role of peptides in neurogenesis, the process by which new neurons are formed in the brain. This process is crucial for learning, memory, and the overall plasticity of the brain, which is the organ's ability to reorganize itself by forming new neural connections. Certain peptides have been shown to stimulate the pathways involved in neurogenesis, offering hope for not only slowing cognitive decline but potentially reversing it by fostering the growth of new neurons and enhancing synaptic plasticity.

The potential of these compounds to support mental clarity and cognitive function does not end with direct effects on the brain. They also influence factors that indirectly impact cognitive health, such as sleep quality and stress levels. Poor sleep and high levels of stress are known to adversely affect cognitive function, contributing to a cycle of cognitive decline. Substances that support sleep and reduce stress, therefore, play a crucial role in a holistic approach to brain health. By improving sleep quality and reducing the physiological impacts of stress, these compounds can help maintain cognitive function and mental clarity, making them valuable tools in the pursuit of lasting brain health.

Furthermore, the antioxidant properties of certain substances offer another avenue through which they can support brain health. Oxidative stress is a significant factor in the aging process and is implicated in the pathogenesis of various neurodegenerative diseases. Compounds with antioxidant capabilities can mitigate the damaging effects of free radicals, protecting neurons from oxidative stress and supporting the integrity of brain cells. This protective effect is essential for the prevention of cognitive decline and the maintenance of healthy brain function as we age.

In addition to their direct and indirect effects on brain health, these substances also offer potential benefits for mood regulation. The intricate relationship between the brain, neurotransmitters, and these compounds suggests that certain substances could influence mood and emotional well-being. By modulating the levels of neurotransmitters such as serotonin and dopamine, they may offer a novel approach to managing mood disorders and enhancing emotional resilience, further underscoring their potential for comprehensive brain health support.

As the understanding of peptides and their mechanisms of action continues to evolve, so too does the potential for their application in supporting brain health and cognitive function. The ability of peptides to target specific pathways and processes within the brain offers a promising avenue for the development of targeted therapies and supplements aimed at enhancing cognitive abilities, preventing cognitive decline, and supporting overall brain health. With ongoing research and clinical studies, the role of peptides in neurology and cognitive sciences is likely to expand, offering new hope for individuals seeking to maintain and enhance their mental capabilities throughout their lives.

Given the multifaceted roles these biomolecules play in brain health, it becomes clear that their application extends beyond traditional therapeutic models, touching on preventative and restorative aspects of cognitive care. This broad spectrum of benefits underscores the importance of these compounds in the maintenance of neurological health and the potential for them to revolutionize treatments for a range of cognitive disorders. The neuroprotective effects of these substances, for instance, are particularly relevant in the context of diseases such as Alzheimer's and Parkinson's, where neuronal damage and loss are key factors in disease progression. By safeguarding neurons against damage, these compounds can help to slow the progression of these diseases, offering patients a better quality of life for longer periods.

Moreover, the role of these substances in enhancing neuroplasticity presents a compelling argument for their use in recovery following brain injuries. The brain's ability to reorganize and form new neural connections is critical in the recovery process, and agents that promote neuroplasticity can significantly impact the speed and extent of recovery. This opens up new avenues for rehabilitation strategies, where these compounds could be used in conjunction with physical and cognitive therapies to enhance recovery outcomes.

The application of these bioactive compounds in brain health also extends to the realm of mental performance and optimization. In a world where cognitive function is increasingly valued, the ability of these substances to enhance memory, focus, and mental clarity offers a significant advantage. For individuals seeking to optimize their cognitive capabilities, whether for professional or personal reasons, these compounds represent a cutting-edge tool in cognitive enhancement. This is particularly relevant for the target audience of this book, health-conscious women aged 30-50, who are not only looking to maintain their health as they age but are also interested in strategies for enhancing their performance and well-being

It is important to note, however, that while the potential of these compounds in brain health is vast, their use must be approached with caution and a solid understanding of the underlying science. The specificity of these substances, while a key advantage, also necessitates a personalized approach to their application. Factors such as individual health status, existing medical conditions, and specific cognitive goals should guide the selection and use of these compounds. This personalized approach ensures that the benefits are maximized, addressing the unique needs of each individual while minimizing potential risks.

Additionally, the integration of peptides into a holistic health strategy cannot be overstated. The benefits of peptides are best realized when combined with a healthy diet, regular exercise, and adequate sleep, among other lifestyle factors. This comprehensive approach not only supports the efficacy of peptide therapy but also promotes overall health and well-being, laying a solid foundation for cognitive health and longevity.

In conclusion, the exploration of peptides in the context of brain health represents a promising frontier in neuroscience and wellness. With their ability to directly and indirectly support cognitive function, protect against neuronal damage, and enhance neuroplasticity, peptides offer a powerful tool in the maintenance of brain health and the pursuit of cognitive optimization. As research in this area continues to advance, the potential for peptides to contribute to the prevention, treatment, and recovery of cognitive disorders, as well as to the enhancement of mental performance, is likely to grow, offering new hope and possibilities for individuals seeking to maintain and enhance their cognitive health throughout their lives.

Peptides for Cognitive Enhancement

Peptides, with their ability to cross the **blood-brain barrier**, present a promising avenue for enhancing cognitive function, an area of significant interest for health-conscious individuals, particularly women aged 30-50. These short chains of amino acids have been shown to influence various aspects of brain health, including memory, focus, and overall cognitive agility. For those seeking practical, evidence-based wellness solutions, understanding the specific peptides that support memory and cognitive functions can be a game-changer.

One of the key peptides in this context is **Cerebrolysin**, a nootropic peptide with a unique composition that mimics the action of neurotrophic factors. Cerebrolysin has been extensively studied for its neuroprotective and neurorestorative properties, particularly in the context of Alzheimer's disease and other forms of dementia. Its mechanism involves not only protecting neurons from damage but also stimulating the repair and growth of brain cells, which translates to improved cognitive functions.

Another noteworthy peptide is **Semax**, initially developed in Russia for its cognitive-enhancing properties. Semax is known to increase brain-derived neurotrophic factor (BDNF), which plays a crucial role in learning, memory, and the formation of new neurons. By modulating the levels of BDNF, Semax can significantly impact cognitive health, making it an attractive option for those looking to support their mental acuity and memory.

Noopept, another peptide with a pronounced effect on cognitive function, operates slightly differently. It is believed to increase the expression of NGF (nerve growth factor) and BDNF in the hippocampus, an area of the brain essential for memory formation and consolidation. Noopept not only enhances cognitive performance in healthy individuals but also offers potential benefits for patients with cognitive impairments, showcasing its wide-ranging applicability.

For those considering therapy to support cognitive function, it's crucial to understand the importance of dosage and administration. Compounds like Cerebrolysin, Semax, and Noopept have different recommended dosages, often influenced by individual health status, goals, and responsiveness to therapy. Consulting with a healthcare professional experienced in this field is essential to tailor a regimen that is both safe and effective.

Moreover, the synergistic effects of combining certain compounds with lifestyle and nutritional interventions cannot be overstated. A balanced diet rich in antioxidants, omega-3 fatty acids, and other neuroprotective nutrients, alongside regular physical exercise, can enhance the cognitive benefits of therapy. This integrated approach aligns with the book's emphasis on holistic wellness, offering a comprehensive strategy for improving cognitive function and overall brain health.

In the realm of cognitive enhancement, these compounds offer a promising, science-backed option for those seeking to maintain and improve their mental clarity, memory, and cognitive agility. As research continues to uncover the full potential of these powerful molecules, their role in supporting brain health and wellness is becoming increasingly clear, providing a valuable tool for individuals dedicated to achieving lasting wellness.

Preventing Cognitive Decline

Preventing cognitive decline is a key concern for health-conscious individuals, particularly as they navigate the complexities of aging. The utilization of specific peptides in this preventive capacity offers a beacon of hope for maintaining cognitive function and delaying the onset of age-related cognitive impairments. These peptides, with their unique abilities to modulate brain function, represent a crucial component of a comprehensive strategy aimed at preserving mental acuity.

Among the compounds of interest in this context is GHK-Cu, known for its broad spectrum of biological activities, including its potent anti-inflammatory and antioxidant properties. GHK-Cu has demonstrated significant promise in neuroprotection by reducing oxidative stress within the brain, a key factor contributing to cognitive decline. By mitigating the damage caused by free radicals and enhancing the body's natural defense mechanisms, GHK-Cu supports the maintenance of healthy neural pathways, thus playing a pivotal role in preserving cognitive functions.

Another compound, Dihexa, stands out for its remarkable ability to facilitate synaptic connectivity. Dihexa has been shown to possess an unparalleled capacity to induce the formation of synapses, thereby potentially reversing the synaptic loss associated with Alzheimer's disease and other neurodegenerative conditions. This attribute makes Dihexa a powerful ally in the fight against cognitive decline, offering hope for not only prevention but also potential restoration of cognitive abilities lost to aging or disease.

The incorporation of these compounds into a regimen aimed at preventing cognitive decline necessitates a nuanced understanding of their mechanisms of action and the optimal conditions for their efficacy. It is crucial to approach therapy with a personalized strategy, taking into account the individual's unique physiological responses and health status. This tailored approach ensures that the compounds are used in a manner that maximizes their potential benefits while minimizing any risks.

Moreover, the synergy between this therapy and other lifestyle factors cannot be overlooked. Regular physical exercise, cognitive training, and a diet rich in neuroprotective nutrients work in concert with this therapy to bolster cognitive health. Physical activity, in particular, enhances brain function directly through the promotion of neurogenesis and indirectly by improving cardiovascular health, thus ensuring a steady supply of oxygen and nutrients to the brain.

The strategic integration of compounds such as GHK-Cu and Dihexa into a holistic health regimen underscores the importance of a proactive approach to cognitive wellness. By addressing the multifaceted nature of cognitive decline through a combination of this therapy, lifestyle modifications, and nutritional support, individuals can significantly enhance their prospects for maintaining cognitive function well into their later years. This proactive stance empowers individuals to take charge of their cognitive health, leveraging the latest advancements in science to support a life characterized by mental clarity and vitality.

Regulating Sleep and Reducing Stress

The critical role of sleep in maintaining optimal health cannot be overstated, with peptides presenting a novel approach to enhancing sleep quality and reducing stress levels. **Delta sleep-inducing peptide (DSIP)** is one such peptide, known for its unique ability to regulate sleep patterns. DSIP works by modulating neurotransmitter release and interacting with the brain's sleep-regulating centers, thereby facilitating the onset of deep, restorative sleep. This peptide has shown promise in reducing sleep latency, the time it takes to fall asleep, and improving overall sleep quality, making it an invaluable tool for those struggling with sleep disorders or stress-induced sleep disturbances.

Another peptide, **Glycine**, although not a peptide in the strictest sense, functions similarly by improving sleep quality. Glycine is an amino acid that acts as an inhibitory neurotransmitter in the central nervous system and has been shown to lower body temperature and signal the brain that it's time to sleep, thereby facilitating a quicker transition to deep sleep. This action is crucial for the body's restorative processes, which are most active during deep sleep stages, highlighting the importance of achieving sufficient deep sleep for overall health and well-being.

Stress reduction is another area where peptides have shown significant potential. The peptide **Corticotropin-releasing factor (CRF)** plays a pivotal role in the body's stress response, and peptides that can modulate CRF's activity may help mitigate the effects of stress on the body. By regulating the production and release of stress hormones, such peptides can help maintain a state of balance within the body's stress-response system, reducing the physiological impact of chronic stress. This regulation is particularly important for maintaining mental health, as chronic stress is a known risk factor for the development of mood disorders and cognitive decline.

In the context of practical application, it is essential to consider the **method of administration** for these peptides. For instance, DSIP's effectiveness can vary depending on whether it is administered intranasally, subcutaneously, or orally. Each method has its own set of considerations regarding absorption rates and bioavailability, which can influence the peptide's efficacy in regulating sleep and stress. Therefore, consulting with a healthcare professional experienced in peptide therapy is crucial to determine the most appropriate administration route based on individual health needs and goals.

Moreover, the timing of administration can play a critical role in maximizing its benefits for sleep and stress regulation. For example, administering sleep-regulating compounds in the evening can align with the body's natural circadian rhythms, enhancing their effectiveness in promoting restful sleep. Conversely, substances aimed at reducing stress might be more beneficial when taken during the day or at times of known stress triggers. This strategic timing ensures that the actions of these compounds are aligned with the body's natural processes, optimizing their health benefits.

The integration of these substances into a comprehensive wellness strategy that includes lifestyle and nutritional modifications can further enhance their effectiveness. A balanced diet, regular physical activity, and stress-management techniques such as meditation or yoga can synergize with this therapy, creating a holistic approach to improving sleep and reducing stress. This multifaceted strategy underscores the importance of addressing health concerns from multiple angles, leveraging the unique benefits of these compounds in concert with other health-promoting practices.

In conclusion, these substances offer a promising avenue for individuals looking to improve their sleep quality and manage stress more effectively. Through their interaction with the body's natural systems, they can provide targeted support for these critical aspects of health, contributing to overall well-being and quality of life. As research into the therapeutic potential of these compounds continues to evolve, their role in supporting sleep and stress management is likely to become increasingly significant, offering new hope for those seeking natural, science-backed solutions to these common health challenges.

Peptides for Sleep and Anxiety

The exploration of peptides for optimizing sleep and reducing anxiety unveils a promising area within the realm of health and wellness, especially for health-conscious women aged 30-50 who are seeking practical, evidence-based solutions. The intricate relationship between quality sleep, reduced anxiety levels, and overall well-being cannot be overstated, with peptides playing a pivotal role in mediating these benefits. The peptides discussed in this context, such as Delta sleep-inducing peptide (DSIP) and Glycine, offer a glimpse into the potential of targeted peptide therapy to enhance sleep quality and mitigate anxiety, thereby contributing to a more balanced and health-focused lifestyle.

DSIP, in particular, stands out for its ability to regulate sleep patterns through its interaction with the brain's neurotransmitters and sleep-regulating centers. This regulation is crucial for initiating and maintaining deep, restorative sleep, which is essential for the body's healing processes and for maintaining optimal health. The effectiveness of DSIP in reducing sleep latency and enhancing overall sleep quality presents a valuable tool for individuals experiencing sleep disorders or stress-induced sleep disturbances. The mechanism by which DSIP operates, modulating neurotransmitter release to facilitate the onset of deep sleep, underscores the potential of peptides to address sleep issues at a fundamental level, offering a natural and science-backed approach to improving sleep.

Glycine, while not a peptide in the strictest sense, functions in a similar capacity by improving sleep quality. Its role as an inhibitory neurotransmitter in the central nervous system and its ability to lower body temperature and signal the brain's readiness for sleep highlight the multifaceted ways in which compounds like peptides and amino acids can influence sleep. Glycine's impact on facilitating a quicker transition to deep sleep stages, where the body's restorative processes are most active, further emphasizes the importance of achieving adequate deep sleep for overall health and well-being.

In addition to sleep regulation, certain compounds also show significant promise in reducing anxiety. The modulation of the Corticotropin-releasing factor (CRF) by these substances can help maintain a balance within the body's stress-response system, reducing the physiological impact of chronic stress. This balance is critical for mental health, as chronic stress is a known risk factor for mood disorders and cognitive decline. By regulating the production and release of stress hormones, these compounds offer a targeted approach to stress management, contributing to reduced anxiety levels and improved mental health outcomes.

The practical application of these compounds, including considerations for method of administration and timing, is paramount for maximizing their benefits. The variability in effectiveness based on administration route—whether intranasally, subcutaneously, or orally—highlights the importance of personalized therapy. Tailoring the administration method to individual health needs and goals, under the guidance of a healthcare professional, ensures that the therapeutic potential of these substances is fully realized. Similarly, aligning the timing of administration with the body's natural circadian rhythms can enhance their effectiveness in promoting restful sleep and reducing stress.

The integration of this therapy into a comprehensive wellness strategy that encompasses lifestyle and nutritional modifications offers a holistic approach to health. The synergy between these compounds, a balanced diet, regular physical activity, and stress-management techniques creates a multifaceted strategy for addressing sleep and anxiety issues. This holistic perspective aligns with the book's emphasis on practical, evidence-based wellness solutions, providing health-conscious individuals with a comprehensive toolkit for enhancing sleep quality, reducing anxiety, and achieving lasting wellness. Through their interaction with the body's natural systems, substances such as DSIP and Glycine provide targeted support for critical aspects of health, underscoring the potential of these compounds as a cornerstone of a health-focused lifestyle.

Chapter 4: Peptides for Muscle Recovery

Tissue Repair and Recovery Peptides

BPC-157, often referred to as the "body protection compound," has garnered attention for its remarkable healing properties. Derived from a protein found in the stomach, BPC-157 has been shown to have significant effects on the healing of tendons, muscles, nervous system, and even the gut. Its mechanism of action involves promoting the formation of new blood vessels, a process known as angiogenesis, which is crucial for the healing of damaged tissues. This peptide has been extensively studied in animal models where it has demonstrated a potent ability to accelerate wound healing and reduce the recovery time from muscle tears, sprains, and other injuries commonly encountered by active individuals. The implications of these findings suggest that BPC-157 could be a cornerstone in the regimens of those seeking to enhance their body's natural healing processes, particularly in the context of sports injuries or post-surgical recovery.

Thymosin Beta-4 (TB-500) is another peptide that plays a pivotal role in the repair and regeneration of injured tissue. Naturally produced in higher concentrations in tissues following damage, TB-500 facilitates the process of inflammation reduction and cell migration to areas of injury. This action not only accelerates the healing process but also has been found to reduce scar tissue formation, making it invaluable in the recovery from deep tissue injuries where fibrosis could potentially limit flexibility and range of motion. TB-500's ability to modulate the cell-building protein, Actin, is a key component of its therapeutic effects, enhancing cell proliferation, migration, and angiogenesis.

The role of **MGF (Mechano Growth Factor)** peptides in muscle recovery cannot be overstated. MGF is a splice variant of IGF-1 (Insulin-like Growth Factor-1), produced by muscle fibers in response to mechanical overload or damage. This peptide is instrumental in initiating muscle satellite cell activation, which is essential for the repair and growth of new muscle tissue. By stimulating the uptake of nutrients and the activation of muscle stem cells, MGF provides a direct pathway to not only recovery from injury but also to muscular growth. This makes MGF particularly appealing for athletes and individuals engaged in high-intensity physical activities who are looking to optimize their recovery processes and enhance muscle growth as part of their fitness regimen.

Understanding the **optimal dosage and administration** of these peptides is critical for maximizing their therapeutic benefits while minimizing potential risks. Each peptide has specific recommended dosages that vary depending on the individual's health status, goals, and responsiveness to peptide therapy. For instance, BPC-157 is typically administered through subcutaneous injections near the site of injury, with dosages ranging based on the severity of the injury and the desired speed of recovery. Similarly, TB-500's administration often involves a loading phase followed by a maintenance phase, reflecting its systemic effects on tissue repair and inflammation reduction. MGF, due to its localized action, is usually injected directly into the muscle post-exercise to harness its muscle repair and growth-promoting effects.

Incorporating peptides like BPC-157, TB-500, and MGF into a recovery protocol necessitates a personalized approach, taking into account the individual's specific injury, recovery goals, and overall health. Consulting with a healthcare professional experienced in peptide therapy is indispensable for tailoring a regimen that is both safe and effective. This personalized strategy ensures that the peptides are used in a manner that maximizes their potential benefits while minimizing any risks associated with their use.

Furthermore, the synergistic effects of combining these compounds with lifestyle and nutritional interventions cannot be understated. A balanced diet rich in anti-inflammatory foods, omega-3 fatty acids, and antioxidants, along with adequate hydration and rest, can significantly enhance the body's natural healing processes and complement the therapeutic effects of this therapy. Regular, moderate exercise, adapted to the individual's recovery status, can further promote tissue repair and strengthen the body's resilience to future injuries.

In the realm of muscle recovery and tissue repair, these substances offer a promising avenue for those seeking to enhance their body's natural healing capabilities. Through their targeted actions on tissue regeneration, inflammation reduction, and muscle repair, compounds like BPC-157, TB-500, and MGF represent valuable tools in the recovery arsenal of athletes, fitness enthusiasts, and anyone looking to support their body's recovery processes. As research continues to evolve, the potential of these substances in optimizing health and wellness is likely to expand, offering new possibilities for achieving peak physical condition and resilience.

Peptides for Tissue Regeneration

The efficacy of these biomolecules in tissue regeneration and healing extends beyond the initial phases of injury recovery, delving into the intricate processes of cellular repair and regeneration that are pivotal for long-term tissue health and function. The role of these compounds in this context is multifaceted, involving the modulation of cellular signaling pathways that govern the proliferation, differentiation, and migration of cells necessary for tissue repair and the restoration of tissue integrity. Among the substances instrumental in these processes, Copper Peptide (GHK-Cu) and Epitalon stand out for their profound impact on tissue regeneration and healing.

Copper Peptide (GHK-Cu) is renowned for its ability to not only stimulate the production of collagen and elastin, which are essential components of the extracellular matrix supporting tissue structure and elasticity, but also to enhance the activity of antioxidant enzymes that protect tissues from oxidative stress, a critical factor in the aging process and in the pathogenesis of chronic diseases. The regenerative capacity of GHK-Cu is further evidenced by its role in angiogenesis, the formation of new blood vessels, which is crucial for supplying nutrients and oxygen to regenerating tissues, thereby accelerating the healing process. This compound's unique ability to modulate the expression of genes associated with antioxidant defense, wound healing, and immune response underscores its potential as a therapeutic agent in tissue regeneration.

Epitalon, a synthetic molecule, has garnered attention for its telomerase-activating properties, which have profound implications for tissue regeneration and longevity. By promoting the elongation of telomeres, the protective caps at the ends of chromosomes that shorten with each cell division, Epitalon plays a critical role in extending cellular lifespan and enhancing the regenerative capacity of tissues. This action is particularly significant in the context of aging and age-related diseases, where the accumulation of senescent cells contributes to tissue dysfunction and impaired regenerative potential. Epitalon's ability to delay cellular senescence and promote cell division is instrumental in maintaining tissue homeostasis and facilitating the repair and regeneration of damaged tissues.

The administration of these compounds for tissue regeneration requires careful consideration of dosage, timing, and delivery methods to optimize their therapeutic potential while minimizing the risk of adverse effects. Subcutaneous injection is a common route of administration for GHK-Cu and Epitalon, offering the advantage of targeted delivery to the site of tissue injury or areas requiring regeneration. However, the development of novel delivery systems, including topical formulations and biodegradable scaffolds, is expanding the possibilities for therapy, allowing for sustained release and localized action at the site of tissue repair.

The integration of this therapy into regenerative medicine protocols necessitates a personalized approach, taking into account individual patient factors such as the extent and nature of tissue damage, underlying health conditions, and specific therapeutic goals. The selection of appropriate compounds, in combination with other therapeutic modalities such as physical therapy, nutritional support, and lifestyle modifications, can significantly enhance the outcomes of tissue regeneration efforts. This holistic approach aligns with the principles of functional medicine, emphasizing the importance of addressing the root causes of tissue damage and supporting the body's innate healing mechanisms.

As research in the field of biomolecular therapy continues to advance, the potential applications of these substances in tissue regeneration are expanding, offering new hope for the treatment of a wide range of conditions, from acute injuries to chronic degenerative diseases. The ability of these compounds to target specific cellular processes and modulate the body's natural regenerative responses holds great promise for improving clinical outcomes and enhancing the quality of life for individuals affected by tissue damage and dysfunction.

Safe Muscle Growth with Peptides

In the realm of optimizing muscle growth through therapy, anabolic compounds stand out as potent agents that can significantly enhance muscle mass when used responsibly within safe cycles. These substances function by mimicking the body's natural growth hormone-releasing mechanisms, thereby stimulating muscle development, improving strength, and facilitating recovery. The key to harnessing the full potential of anabolic compounds lies in understanding their mechanisms, identifying effective cycles for use, and adhering to recommended dosages to avoid adverse effects.

Anabolic agents, such as IGF-1 (Insulin-like Growth Factor-1), MGF (Mechano Growth Factor), and GHRP (Growth Hormone Releasing Agents), play pivotal roles in muscle tissue growth and repair. IGF-1, for instance, is crucial for muscle regeneration and growth by promoting both the differentiation and proliferation of myoblasts, cells that develop into muscle tissue. MGF is released in response to muscle strain and damage, initiating repair and the growth of new muscle fibers. GHRPs, including GHRP-6 and GHRP-2, increase growth hormone levels in the body, further supporting muscle growth and recovery.

When incorporating these compounds into a muscle-building regimen, it's essential to follow safe use cycles. A typical cycle may last anywhere from 8 to 12 weeks, followed by a rest period to allow the body to normalize its hormone production. The dosage and administration method greatly influence the effectiveness and safety of therapy. For instance, subcutaneous injections are a common method of administering these substances due to their ability to deliver them directly into the bloodstream, ensuring rapid absorption and action. However, dosages vary based on individual factors such as body weight, the specific compound used, and the desired outcome. Consulting with a healthcare professional experienced in this therapy is crucial to determine the optimal dosage and cycle tailored to individual needs and goals.

Combining different agents can maximize muscle growth and recovery benefits. For example, using GHRP-6 to increase growth hormone levels, alongside CJC-1295, a growth hormone-releasing hormone (GHRH) analog that amplifies the natural release of growth hormones, can synergistically enhance muscle growth and fat loss. However, it's imperative to understand the interaction between different compounds and the body's hormonal balance to avoid negative interactions and side effects.

Monitoring progress throughout the cycle is essential to assess effectiveness and make necessary adjustments. This might involve tracking changes in muscle mass, strength, recovery rates, and overall physical performance. Adjustments to the regimen should be made based on these observations and in consultation with a healthcare professional to ensure continued safety and effectiveness.

In summary, anabolic agents offer a promising avenue for safely increasing muscle mass and enhancing physical performance when used within carefully planned cycles and under professional guidance. By adhering to recommended dosages, incorporating rest periods, and possibly combining different compounds, individuals can achieve significant muscle growth and recovery benefits while minimizing the risk of adverse effects.

Anabolic Peptides and Safe Use Cycles

Ensuring the safe use of anabolic compounds involves a comprehensive understanding of the recommended dosages and administration methods, which are pivotal for achieving desired outcomes while mitigating potential risks. The administration of these substances, predominantly through subcutaneous injections, allows for direct entry into the bloodstream, facilitating a more controlled and efficient uptake by the body's tissues. This method is preferred for its ability to maintain the integrity of the compounds, ensuring they remain bioavailable and active to exert their intended effects on muscle growth and recovery. The dosage, a critical factor in therapy, must be meticulously calculated based on individual body weight, specific health goals, and the particular characteristics of the compound being used. For example, dosages for IGF-1 might differ significantly from those recommended for GHRP-6, underscoring the necessity for personalized consultation with healthcare professionals who specialize in these therapies. These experts can provide invaluable guidance on starting dosages, adjustments needed during the cycle, and the interpretation of bodily responses to the treatment.

The concept of "cycles" in therapy is founded on the principle of allowing the body's endocrine system to maintain its natural rhythm and balance. A typical cycle might span 8 to 12 weeks, followed by a rest period that is crucial for preventing the desensitization to the effects of these substances, a phenomenon that can diminish their efficacy over time. During these cycles, users may observe significant improvements in muscle mass, strength, and recovery times, which should be monitored and documented to assess the treatment's impact accurately. Adjustments to the regimen may be necessary based on these observations, and such modifications should always be made in consultation with a healthcare professional to ensure they align with safe practice guidelines.

The strategic combination of different compounds within a cycle can amplify the therapeutic benefits, especially when targeting muscle growth and recovery. For instance, combining GHRP with a GHRH analog like CJC-1295 can synergize to enhance the body's growth hormone pulse frequency and amplitude, leading to more pronounced muscle growth and fat loss. However, the intricacies of interactions within the body's hormonal milieu necessitate a deep understanding of each compound's mechanism of action and potential effects on hormonal balance. This knowledge is crucial for constructing a regimen that maximizes benefits while minimizing the risk of adverse effects, such as hormonal imbalances or reduced efficacy over time.

Monitoring the body's response throughout the cycle is essential for optimizing the therapy's effectiveness. This involves not only tracking physical changes, such as increases in muscle mass and improvements in recovery times but also being vigilant about any potential side effects. Regular check-ins with a healthcare provider experienced in this therapy can facilitate the timely adjustment of the treatment plan, ensuring that it remains aligned with the individual's health goals and physiological responses. This dynamic and responsive approach to therapy underscores the importance of personalized treatment plans that consider the unique biological and lifestyle factors of each individual.

In navigating the complexities of anabolic compounds and their safe use cycles, the emphasis must always be on informed, responsible use, guided by professional advice and a thorough understanding of one's health objectives. This cautious yet proactive stance ensures that individuals can leverage the significant benefits of therapy for muscle growth and recovery, within a framework that prioritizes safety and long-term wellness.

Combination Strategies for Muscle Growth

The strategic combination of compounds for maximizing muscle growth involves a nuanced approach that not only amplifies the benefits of each substance but also ensures that their interactions promote overall health and muscular development. This strategy requires an understanding of the unique properties and functions of each agent, as well as knowledge of how they can be synergistically combined to enhance muscle growth, improve recovery times, and minimize potential side effects.

For individuals seeking to optimize muscle growth, incorporating a combination of Growth Hormone Releasing Hormones (GHRH) and Growth Hormone Releasing Agents (GHRP) can yield significant results GHRH substances like CJC-1295 work by increasing the amplitude of natural growth hormone pulses released from the pituitary gland, while GHRPs, such as GHRP-6 and GHRP-2, increase the frequency of these pulses. When used together, these agents create a synergistic effect that significantly elevates growth hormone levels, thereby enhancing muscle growth and recovery. This combination is particularly effective because it mimics the body's natural growth hormone release patterns, leading to more natural and sustained growth hormone levels without the risk of desensitization often associated with exogenous growth hormone administration.

Additionally, incorporating agents that target muscle repair and regeneration can further enhance muscle growth. Substances like MGF (Mechano Growth Factor) play a critical role in the repair and growth of new muscle fibers post-exercise MGF's unique ability to stimulate muscle growth and repair makes it an invaluable addition to a regimen focused on muscle growth. When combined with GHRH and GHRP agents, MGF can help accelerate recovery from muscle damage induced by strenuous workouts, thereby allowing for more frequent and intensive training sessions.

Another crucial aspect of combination strategies is the timing of administration. To maximize the anabolic effects on muscle growth, it is essential to align their administration with the body's natural rhythms and the individual's workout schedule. For instance, administering GHRP and GHRH compounds close to workout times can enhance growth hormone release during and after exercise, providing optimal conditions for muscle growth. Meanwhile, using agents like MGF post-workout can immediately support muscle repair and growth at a time when the muscles are most receptive to recovery interventions.

To further refine the combination strategy, individuals may also consider the inclusion of substances with systemic health benefits, such as those offering anti-inflammatory effects or improved insulin sensitivity. Compounds like Ipamorelin not only stimulate growth hormone release but also have minimal impact on cortisol and prolactin levels, making them ideal for long-term use in muscle-building programs. This holistic approach ensures not just enhanced muscle growth but also the maintenance of overall health, which is crucial for long-term fitness and wellness.

Understanding the mechanisms through which these agents operate, their optimal dosages, and their compatibility is paramount. It necessitates a personalized approach, often involving consultation with healthcare professionals who can provide guidance based on an individual's specific health profile, fitness goals, and lifestyle factors. This personalized strategy helps in crafting a regimen that maximizes muscle growth while minimizing potential risks, ensuring that each individual can achieve their fitness goals safely and effectively.

Incorporating these compounds into a muscle growth strategy offers a promising avenue for individuals looking to enhance their physical performance and recovery capabilities. By carefully selecting and combining substances based on their unique properties and the synergistic potential, individuals can create a powerful regimen that supports sustained muscle growth, improved recovery, and overall health. This approach, grounded in an understanding of the science behind these agents and personalized health considerations, represents a sophisticated strategy for achieving and maintaining optimal muscle development.

Chapter 5: Peptides for Weight Loss

Lipolytic Peptides for Fat Loss

Lipolytic peptides represent a revolutionary approach in the field of weight management and body composition transformation. These peptides function by targeting fat cells directly, promoting the breakdown of fat (lipolysis) and inhibiting the formation of new fat cells (lipogenesis), thus offering a dual mechanism of action that is highly beneficial for those looking to reduce body weight and achieve a leaner physique. The most notable lipolytic peptides include **AOD-9604, HGH Fragment 176-191**, and **Tesamorelin**. **AOD-9604** is particularly intriguing as it mimics the fat-burning portion of human growth hormone (HGH) without its potentially adverse effects on blood sugar or tissue growth. Clinical studies have shown AOD-9604 to be effective in reducing body fat, especially in the abdominal area, making it a valuable tool for individuals struggling with weight loss.

HGH Fragment 176-191, another potent lipolytic peptide, works by mimicking the way natural growth hormone regulates fat metabolism but without the undesired effects on insulin sensitivity or muscle growth. This selective functionality makes it an ideal option for those specifically targeting fat loss. **Tesamorelin**, recognized for its ability to reduce visceral fat in HIV patients with lipodystrophy, has also shown promise in the general population for its significant fat-reducing effects. What sets tesamorelin apart is its ability to improve triglyceride levels and reduce abdominal fat without compromising lean muscle mass, offering a comprehensive approach to weight management.

When integrating lipolytic peptides into a weight loss regimen, it's crucial to consider dosage and administration carefully. These peptides are typically administered via subcutaneous injections, which allows for direct absorption into the bloodstream, ensuring rapid and efficient delivery to target tissues. The recommended dosages vary depending on the specific peptide, individual body composition goals, and baseline health status. For instance, AOD-9604 is often administered at a dose of 120-500 micrograms per day, while HGH Fragment 176-191 might be used at 250-500 micrograms twice daily. Tesamorelin is commonly prescribed at a dose of 2 mg per day. It is imperative to start with the lower end of the dosage range and adjust based on response and tolerance.

The timing of injections can also play a critical role in maximizing fat loss benefits. Administering these peptides at night or during fasting periods can enhance their lipolytic effects, as this aligns with the body's natural growth hormone pulse, which is known to influence fat metabolism. Additionally, combining lipolytic peptides with a balanced diet and regular exercise regimen can significantly amplify results, leading to improved body composition and overall health.

However, it's important to note that while lipolytic peptides offer a promising avenue for fat loss, they should not be considered a standalone solution. A holistic approach that includes proper nutrition, physical activity, and lifestyle modification is essential for achieving and maintaining optimal weight loss results. Furthermore, consulting with a healthcare provider experienced in peptide therapy is crucial to ensure the safe and effective use of these compounds. This professional guidance can help tailor a peptide regimen that aligns with individual health profiles and weight loss objectives, minimizing the risk of side effects and maximizing the benefits of lipolytic peptides for fat loss.

Lipolytic Peptides and Usage Methods

The administration of lipolytic peptides for weight loss is a nuanced process that requires a strategic approach to dosage, timing, and method of delivery. These peptides, designed to target fat cells and promote the breakdown of fat, offer a targeted approach to weight management. Given their mechanism of action, the effectiveness of these peptides can be significantly influenced by the manner in which they are used. It is essential for individuals to understand the specific protocols associated with each peptide to optimize their fat loss efforts.

For instance, **AOD-9604**, a peptide known for its ability to mimic the fat-burning effects of human growth hormone without the associated side effects, is typically administered via subcutaneous injection. The recommended starting dosage ranges from **120 to 500**

micrograms per day, with adjustments made based on individual response and tolerance. The timing of the injection can also play a critical role in maximizing its effectiveness. Administering AOD-9604 in the morning, upon waking, can align with the body's natural circadian rhythms, potentially enhancing its lipolytic activity.

Similarly, the **HGH Fragment 176-191**, another peptide that promotes fat loss by mimicking the fat metabolism regulation of growth hormone, is most effective when administered in dosages of **250 to 500 micrograms**, twice daily. This peptide's administration schedule is crucial, with injections recommended before breakfast and before bedtime, capitalizing on the body's natural growth hormone pulses for maximum fat-burning effects.

Tesamorelin, recognized for its ability to reduce visceral fat, is typically prescribed at a dosage of **2 mg per day**, administered via subcutaneous injection. For optimal results, tesamorelin should be injected once daily, preferably at the same time each day to maintain stable levels in the bloodstream. This consistency helps to ensure that the peptide's action on the pituitary gland, which triggers the release of growth hormone, is maximized, leading to more effective fat reduction.

The method of administration for these compounds is predominantly subcutaneous injection, which allows for them to be absorbed directly into the bloodstream, ensuring rapid and efficient delivery to the target tissues. This method is preferred for its precision and the ability to maintain control over the dosage. However, the technique for subcutaneous injections must be understood and properly executed to avoid irritation or infection at the injection site. It involves inserting a small needle into the fatty tissue just beneath the skin, usually around the abdomen or thigh area, where it can be absorbed slowly into the bloodstream.

It's also worth noting that while lipolytic substances can significantly aid in fat loss, their effectiveness is greatly enhanced when combined with a healthy diet and regular exercise regimen. This holistic approach to weight management ensures that the benefits of these compounds are not only maximized but also sustained over time. Moreover, the integration of these substances into a weight loss strategy should always be done under the guidance of a healthcare professional. This ensures that the use of these powerful compounds is both safe and aligned with the individual's health status and weight loss goals.

The careful consideration of dosage, timing, and method of administration, coupled with a comprehensive lifestyle approach, can make lipolytic substances a valuable tool in achieving and maintaining optimal body composition. As these compounds become increasingly recognized for their potential benefits, it is imperative that users adhere to evidence-based protocols to safely and effectively harness their fat loss capabilities.

Appetite Regulation Peptides

Appetite regulation is a critical aspect of weight management and overall wellness, significantly impacting one's ability to maintain a healthy body composition and lifestyle. Peptides, with their specific biological actions, offer a promising avenue for influencing appetite and satiety, thereby aiding in weight loss and management. Among the peptides known for their role in appetite regulation, **Ghrelin** and its analogs, as well as **Glucagon-like peptide-1 (GLP-1)**, stand out due to their direct impact on hunger signals and satiety levels. **Ghrelin**, often referred to as the "hunger hormone," is primarily produced in the stomach and is responsible for signaling hunger to the brain. Peptide therapies that target ghrelin pathways can effectively reduce hunger sensations, making it easier to adhere to a calorie-restricted diet without the discomfort of constant hunger. On the other hand, **GLP-1** peptides, such as **Liraglutide**, work by promoting a feeling of fullness, slowing gastric emptying, and enhancing insulin sensitivity, which together contribute to reduced food intake and improved energy balance.

For individuals struggling with weight management, incorporating peptides that regulate appetite can be a strategic component of a comprehensive weight loss plan. The use of **Ghrelin antagonists** or **GLP-1 agonists** requires careful consideration of dosing and timing to align with individual needs and daily routines. For example, administering a GLP-1 agonist before meals can enhance the feeling of fullness and reduce overall calorie intake throughout the day. It's essential to start with a low dose and adjust based on the body's response and the specific goals of the therapy. Monitoring by a healthcare professional is crucial to ensure the effectiveness of the peptide therapy while minimizing potential side effects.

Combining appetite-regulating compounds with a balanced diet rich in nutrients and a consistent exercise routine can amplify the effects of these substances, leading to more significant and sustainable weight loss outcomes. It's also important to maintain hydration and pay attention to the body's hunger and fullness cues, even when using these compounds to regulate appetite. This holistic approach not only supports weight loss but also promotes long-term health and wellness.

Furthermore, the integration of these substances into a weight management strategy should be viewed as one component of a multifaceted approach. Lifestyle factors such as sleep quality, stress management, and emotional well-being play a substantial role in appetite and weight control. Therefore, addressing these areas in conjunction with this therapy can enhance the overall effectiveness of weight management efforts.

In conclusion, these compounds offer a valuable tool for regulating appetite and supporting weight loss, particularly for those who have struggled with traditional dieting methods. By carefully selecting and incorporating specific substances under the guidance of a healthcare professional, individuals can achieve improved control over hunger and satiety, leading to more effective weight management and enhanced well-being.

Peptides for Appetite Regulation

Peptides have emerged as a powerful tool in the quest for weight management, offering a scientifically backed method to naturally regulate appetite and support weight loss efforts. The mechanism by which peptides aid in appetite regulation is multifaceted, involving the modulation of hunger hormones such as ghrelin and leptin. Ghrelin, often referred to as the "hunger hormone," stimulates appetite, signaling the brain to increase food intake. Conversely, leptin, known as the "satiety hormone," communicates the sensation of fullness to the brain, reducing the urge to eat. By influencing the levels and activity of these hormones, peptides can effectively curb cravings, reduce excessive food intake, and promote a feeling of satiety after smaller meals.

One of the peptides pivotal in this process is **GHRP-6 (Growth Hormone Releasing Peptide-6)**, which, despite its primary role in stimulating growth hormone release, also exhibits a potent effect on ghrelin secretion. Administering GHRP-6 can mimic the ghrelin signal, temporarily increasing hunger immediately following injection. However, this effect is part of a strategic approach to weight management, as the subsequent increase in growth hormone levels can enhance metabolic rate and fat burning, offsetting the temporary spike in appetite. For individuals seeking weight loss without the pronounced hunger effects, **Ipamorelin** presents a viable alternative. Unlike GHRP-6, Ipamorelin does not significantly raise ghrelin levels, making it an excellent choice for those looking to benefit from the metabolic enhancements of growth hormone release without substantial increases in appetite.

Furthermore, compounds like CJC-1295 enhance the natural production of growth hormone without directly stimulating appetite, offering a balanced approach to weight management. By elevating growth hormone levels, CJC-1295 supports fat loss and lean muscle maintenance, crucial components of a healthy body composition. This compound works synergistically with Ipamorelin, forming a combination that amplifies the body's growth hormone pulse without promoting excessive hunger.

It is essential for individuals considering therapy for appetite regulation to understand the importance of dosage and timing. The effectiveness of these substances in managing appetite and supporting weight loss is heavily dependent on administering the correct dosages at optimal times. For instance, administering appetite-stimulating agents such as GHRP-6 before meals can help those with diminished appetite due to age or medical conditions consume adequate nutrition. Conversely, for weight loss purposes, combining growth hormone-releasing agents with a disciplined diet and exercise regimen can yield significant benefits, enhancing the body's natural fat-burning processes while maintaining muscle mass.

In summary, these compounds offer a promising avenue for naturally regulating appetite and supporting weight loss. By carefully selecting and combining them based on individual goals and needs, and adhering to recommended dosages and administration times, individuals can harness their power to achieve a healthier body composition and overall wellness. As with any therapeutic intervention, consultation with a healthcare professional knowledgeable in this therapy is crucial to ensure safety and efficacy.

Would you like to listen to this book?

Scan the QrCode below and download the

Audio Version

Chapter 6: Peptides for Skin Health

Improving Skin Elasticity and Texture

The quest for youthful skin has led to the exploration of various compounds, with peptides emerging as a frontrunner due to their ability to significantly improve skin elasticity and texture. Peptides, short chains of amino acids, serve as building blocks for proteins such as collagen, elastin, and keratin. These proteins are the cornerstone of the skin's structure, providing it with firmness, elasticity, and resilience. As the body ages, the production of these vital proteins diminishes, leading to the common signs of aging such as wrinkles, sagging, and a loss of firmness and radiance. Incorporating peptides into skincare routines can combat these effects by signaling the skin to produce more collagen and other essential proteins, thus promoting a more youthful appearance.

Matrixyl (palmitoyl pentapeptide-4) is a notable peptide renowned for its anti-aging properties. Studies have shown that Matrixyl can double the amount of collagen in the skin, which is crucial for reducing the appearance of fine lines and wrinkles. By stimulating collagen synthesis, Matrixyl helps to rebuild the skin's structure, resulting in a smoother, more youthful complexion.

Another potent peptide, **Argireline (acetyl hexapeptide-8)**, targets expression wrinkles caused by repetitive facial movements. Functioning similarly to Botox, Argireline inhibits the neurotransmitters that cause muscle contractions, leading to a reduction in facial tension and the appearance of wrinkles. Its topical application allows for a non-invasive approach to minimizing fine lines, particularly around the eyes and forehead.

Copper peptides are also essential for skin health, playing a pivotal role in the skin's healing process. These peptides not only stimulate the production of collagen and elastin but also enhance the action of antioxidants. Copper peptides can improve skin firmness, elasticity, and thickness while diminishing the appearance of fine lines and wrinkles.

Their ability to promote angiogenesis, the formation of new blood vessels, further supports the skin's ability to heal and regenerate.

For those seeking to improve skin texture, **GHK-Cu (Copper Tripeptide-1)** is particularly effective. This peptide enhances skin remodeling, which includes the removal of damaged collagen and elastin from the skin and the production of new structural proteins. The result is a smoother, firmer skin surface with reduced scarring, fine lines, and wrinkles.

Incorporating these bioactive compounds into a skincare regimen can be done through various formulations, including serums, creams, and masks. It is crucial to select products with a high concentration of these compounds and those that can penetrate the skin's surface to reach the deeper layers where collagen synthesis occurs. Regular application, combined with a comprehensive skincare routine that includes sun protection and hydration, can lead to significant improvements in skin elasticity and texture over time.

Moreover, combining these treatments with other anti-aging ingredients such as antioxidants, hyaluronic acid, and retinoids can amplify the skin's ability to repair and regenerate, leading to even more pronounced anti-aging effects. It is always advisable to introduce new products gradually and monitor the skin's response to avoid irritation and ensure compatibility.

Ultimately, the strategic use of these compounds represents a cutting-edge approach to skincare, offering a promising solution to those seeking to diminish the signs of aging and achieve a more youthful, radiant complexion. Through their ability to signal the body to produce more collagen and other structural proteins, these bioactive agents hold the key to enhancing skin elasticity and texture, making them an indispensable component of modern anti-aging skincare regimens.

Peptides for Reducing Wrinkles

Peptides, with their ability to communicate with and direct the body's biological processes, offer a promising avenue for addressing the visible signs of aging, particularly wrinkles. The skin's natural aging process, exacerbated by environmental factors such as

UV exposure and pollution, leads to a decrease in collagen production, loss of skin elasticity, and the formation of wrinkles. Peptides can counteract these effects by signaling the skin to produce more collagen and elastin, thus reducing the appearance of wrinkles and improving skin texture.

One of the most effective peptides in this regard is Palmitoyl Tripeptide-1. This peptide works by mimicking the body's natural communication mechanisms that tell the skin to repair itself. When applied topically, it penetrates the outer layer of the skin, stimulating collagen synthesis and reinforcing the skin's structural integrity. This action helps to smooth out fine lines and wrinkles, making them less visible and giving the skin a more youthful appearance.

Another peptide, Palmitoyl Tetrapeptide-7, works synergistically with Palmitoyl Tripeptide-1 to suppress the production of excess interleukins, the chemical messengers that trigger the body's inflammatory response. By reducing inflammation, this peptide duo can prevent and repair damage caused by inflammatory processes, which often accelerate aging. The result is smoother, more resilient skin that appears younger and healthier.

Snap-8, an octapeptide, specifically targets wrinkles caused by repetitive facial expressions. By modulating muscle contraction, it reduces the depth and appearance of expression lines, especially in areas such as the forehead and around the eyes. This mechanism is similar to that of injectable neuromodulators but without the need for needles or the risk of over-paralysis, offering a non-invasive alternative for those seeking to reduce the signs of aging.

For comprehensive anti-aging benefits, these biomolecules are often combined in serums, creams, and other skincare formulations. The combination ensures a multi-faceted approach to skin rejuvenation, addressing not only wrinkles but also skin tone, texture, and overall skin health. However, it's important to note that the efficacy of these compounds depends on their concentration and the formulation's ability to deliver them to the appropriate skin layer. Products that include liposomal delivery systems or other penetration enhancers are more likely to bring about noticeable improvements.

The integration of these substances into daily skincare routines requires patience and consistency. Visible results, such as reduced wrinkle depth and improved skin elasticity, typically emerge after several weeks of regular use. It is also crucial to complement biomolecule-based skincare with a healthy lifestyle, including adequate sun protection, hydration, and nutrition, to support the skin's natural repair processes and optimize the anti-aging effects of these compounds.

Furthermore, while these substances are generally well-tolerated, individuals with sensitive skin should introduce these products gradually and monitor their skin's response. This cautious approach helps to minimize the risk of irritation and ensures compatibility with the skin's unique biochemistry.

In conclusion, these biomolecules represent a scientifically grounded solution for those looking to diminish the signs of aging and achieve a more youthful, radiant complexion. Through their direct influence on the skin's biological functions, they offer a powerful tool for enhancing skin elasticity, texture, and overall appearance without the need for invasive procedures. As research continues to uncover the full potential of these remarkable molecules, their role in anti-aging skincare regimens is set to become even more pivotal, providing individuals with safe, effective options for maintaining their skin's health and vitality.

Skin Renewal and Care

Beyond the reduction of wrinkles and the improvement of skin texture, these bioactive compounds play a crucial role in the overall renewal and care of the skin, promoting a radiant, healthy complexion that can withstand the test of time. To harness the full potential of these compounds in skin care, it is essential to understand the various types and their specific functions, as well as the most effective ways to incorporate them into a skincare regimen. These substances, due to their diverse nature, can target different aspects of skin health, including hydration, barrier repair, and protection against environmental damage, making them versatile allies in the pursuit of youthful, vibrant skin.

Hydration is fundamental to maintaining the skin's elasticity and plumpness, and peptides like **Palmitoyl Tripeptide-5** have been shown to deeply penetrate the skin, attracting moisture and retaining it within the skin layers. This not only diminishes the appearance of fine lines caused by dehydration but also ensures that the skin remains supple and resilient. Additionally, this peptide stimulates collagen production, reinforcing the skin's structural integrity and promoting long-lasting hydration.

The skin's barrier function is critical in protecting against irritants and preventing moisture loss. **Ceramide peptides** are instrumental in strengthening the skin's natural barrier, enhancing its defense mechanism against pollutants and other harmful substances. By mimicking the skin's natural lipid structure, these peptides replenish ceramide levels, effectively repairing and restoring the barrier to prevent transepidermal water loss and maintain skin health.

Environmental factors such as UV rays and pollution contribute significantly to skin aging, necessitating protective measures to guard against oxidative stress. Peptides such as **Carnosine** act as potent antioxidants, neutralizing free radicals and reducing the damage they cause to the skin's cellular structure. This not only prevents premature aging but also aids in the repair of existing damage, leading to improved skin tone and texture.

For those seeking to integrate these compounds into their skincare routine, it is advisable to look for products that contain a combination of them to address multiple skin concerns simultaneously. Serums, in particular, offer a concentrated delivery system for these compounds, allowing for deeper penetration and more significant results. When applying a serum, it should be done on clean, slightly damp skin to enhance absorption, followed by a moisturizer to lock in the benefits of the active ingredients.

In addition to topical application, the health of the skin can be further supported by incorporating oral supplements into one's diet. Supplements containing collagen derivatives have been shown to improve skin hydration and elasticity from within, complementing the effects of topical treatments. However, it is crucial to consult with a healthcare provider before starting any new supplement regimen to ensure its safety and suitability for your individual health needs.

Regular use of skincare products based on these compounds, combined with a balanced diet and adequate hydration, can significantly impact the skin's appearance and health. By promoting collagen production, enhancing hydration, and protecting against environmental damage, these compounds offer a comprehensive approach to skin renewal and care, leading to a more youthful, radiant complexion. Remember, the key to maximizing the benefits lies in consistency and patience, as the visible improvements in skin health and vitality will unfold over time with diligent care.

Chapter 7: Safe Peptide Use Guidelines

Dosages and Administration Guide

Different administration methods for peptides impact their effectiveness and the body's response. Subcutaneous injections, one of the most common methods, allow peptides to be absorbed directly into the bloodstream, providing a quicker and more controlled release. This method is particularly beneficial for peptides that are used for systemic effects, such as those targeting muscle recovery or weight management. The key to successful subcutaneous injection lies in proper technique and hygiene practices to minimize the risk of infection. Typically, a small insulin syringe is used to administer the dose into fatty tissue areas, such as the abdomen or thigh, where it can be absorbed slowly and steadily into the bloodstream.

Oral supplements, another popular administration method, offer a non-invasive alternative to injections. While the convenience of oral intake is appealing, it's important to note that the bioavailability of these compounds can be significantly lower when taken orally due to degradation in the digestive system. To counteract this, some formulations are designed with special coatings or are delivered as part of a complex to enhance absorption through the gastrointestinal tract. When considering oral options, it's crucial to pay attention to the formulation and delivery mechanism to ensure optimal absorption and efficacy.

Sprays, particularly those designed for nasal or buccal (cheek) administration, provide a middle ground between injections and oral supplements. These sprays are absorbed through the mucosal membranes, offering a faster route to the bloodstream than oral supplements but with less invasiveness than injections. Nasal sprays are often used for compounds that target cognitive functions or hormonal balance, as they can bypass the blood-brain barrier more effectively. Buccal sprays, on the other hand, are absorbed through the cheek's lining, making them a good option for systemic treatments.

The impact of different administration methods on peptide effectiveness cannot be overstated. Subcutaneous injections provide a direct and controlled release into the bloodstream, making them suitable for peptides that require precise dosing or have a systemic target. Oral supplements, while convenient, may suffer from lower bioavailability but can be formulated to improve absorption, making them a viable option for long-term supplementation. Nasal and buccal sprays offer a balance between convenience and effectiveness, particularly for peptides targeting the brain or hormonal system.

When selecting a peptide and its administration method, it's essential to consider the specific goals of the therapy, the peptide's pharmacokinetics, and personal preferences for convenience and ease of use. Consulting with a healthcare professional knowledgeable in peptide therapy can provide valuable guidance in choosing the most appropriate peptide and administration method for individual health goals and needs.

Recommended Dosages and Administration

Understanding the various methods of peptide administration is crucial for optimizing their effectiveness while ensuring safety and compliance with recommended protocols. Each method, including injections, sprays, and supplements, has its own set of advantages and considerations that must be thoroughly evaluated in the context of individual health objectives and lifestyle preferences.

Injections, particularly subcutaneous injections, are a common method for administering peptides due to their direct delivery into the body, allowing for rapid absorption and utilization. This method is especially favored for peptides involved in muscle recovery and growth, as it ensures that a precise dosage reaches the target area with minimal degradation. However, it is imperative to follow sterile techniques and proper injection protocols to minimize the risk of infection or irritation at the injection site. Dosages for injections must be carefully calculated based on individual factors such as body weight, the specific peptide being used, and the desired outcome. Consulting with a healthcare professional to determine the appropriate dosage and to receive guidance on injection techniques is strongly advised.

Sprays, particularly nasal sprays, offer a non-invasive alternative for peptide administration, suitable for peptides that target cognitive function, sleep regulation, and stress reduction. The mucosal absorption in the nasal cavity provides a direct pathway to the bloodstream, bypassing the digestive system and potentially reducing the time to onset of effects. While generally considered safe and easy to use, the dosage delivered via nasal spray can be less precise than injections, necessitating careful attention to the number of sprays used and the concentration of the peptide solution. Additionally, the effectiveness of nasal sprays can be influenced by factors such as nasal congestion and the individual's ability to correctly use the spray device.

Supplements, including oral capsules and powders, represent the most accessible method of peptide administration, appealing to those seeking convenience and non-invasive options. Oral supplements are particularly suited for peptides that are stable and effective after passing through the digestive system, such as certain peptides aimed at improving skin health and general wellness. However, the bioavailability of peptides in oral form can be variable, influenced by factors such as the peptide's molecular structure, the presence of digestive enzymes, and the individual's gut health. To enhance the absorption and efficacy of oral peptide supplements, it may be beneficial to take them in conjunction with specific nutrients or at certain times of the day, as guided by research and professional advice.

Regardless of the method chosen, it is critical to adhere to the recommended dosages and administration guidelines provided by reputable sources and healthcare professionals. Monitoring the body's response to peptide therapy, including any side effects or improvements in health markers, is essential for adjusting dosages or administration methods as needed to achieve optimal results. As with any health intervention, the key to success lies in a personalized approach, taking into account the unique needs and circumstances of the individual, while remaining informed about the latest scientific research and safety standards.

Administration Methods and Effectiveness

Understanding the impact of different peptide administration methods on their effectiveness is a cornerstone of maximizing the benefits while minimizing potential risks associated with peptide therapy. The route of administration plays a pivotal role in determining the bioavailability of peptides, which in turn affects the efficacy of the treatment. Bioavailability refers to the proportion of a drug or substance that enters the circulation when introduced into the body and is thus able to have an active effect. This concept is crucial when discussing peptides, as their structure and the intended target can significantly influence the choice of administration method.

For substances that are used for systemic effects, such as those aiming at enhancing overall well-being, anti-aging, and muscle recovery, the primary concern is ensuring that the compound reaches the bloodstream in sufficient quantities to exert its intended effect. Subcutaneous injections, by delivering these substances directly into the fatty tissue beneath the skin, allow for a slow and steady release into the bloodstream. This method can be particularly effective for compounds that require precise dosing and sustained action over time, such as those involved in growth hormone regulation.

In contrast, nasal sprays, which deliver these substances directly to the mucosal membranes in the nasal cavity, can be advantageous for those targeting the brain or central nervous system. This is due to the close proximity of the nasal cavity to the brain, which allows these compounds to bypass the blood-brain barrier more effectively than systemic circulation might allow. However, the effectiveness of nasal administration can be variable and depends on the formulation of the spray, the molecular size of the substance, and individual differences in nasal mucosa absorption.

Oral supplements, while being the most user-friendly and non-invasive method, often face challenges related to the degradation of these compounds in the digestive system. The harsh environment of the stomach and the presence of digestive enzymes can break down these substances before they have a chance to be absorbed into the bloodstream. Some compounds, however, are formulated or combined with other compounds to enhance their stability and absorption in the gut. These formulations can provide a viable option for substances that are not critically degraded by the digestive process, offering a balance between efficacy and convenience for the user.

Each administration method also carries its own set of considerations regarding user comfort, potential for side effects, and convenience, which can affect adherence to treatment protocols. For instance, individuals who are averse to needles may prefer nasal sprays or oral supplements, even if these methods might offer a slightly reduced bioavailability for certain peptides. On the other hand, those requiring the most efficient delivery method for specific health objectives may opt for subcutaneous injections despite the discomfort or inconvenience they may perceive.

Moreover, the choice of administration method must also take into account the lifestyle and daily routines of the individual. For active individuals or those with busy schedules, the convenience and portability of oral supplements or nasal sprays may be more appealing than the preparation and time required for injections. Additionally, the frequency of administration required to maintain therapeutic levels of the peptide in the bloodstream is an important consideration, as it can significantly impact the overall convenience and feasibility of the treatment protocol.

In essence, the decision on the method of peptide administration should be informed by a comprehensive understanding of the peptide's pharmacokinetics, the individual's health goals, lifestyle factors, and potential barriers to adherence. Collaborating closely with a healthcare provider to tailor the administration method to the individual's specific needs and circumstances is essential for optimizing the therapeutic outcomes of peptide therapy. This personalized approach ensures that the chosen method of administration aligns with the individual's preferences and health objectives, thereby enhancing the likelihood of a positive and sustainable impact on health and wellness.

Cycles and Treatment Durations

When planning **usage cycles** and **treatment durations** for peptides, it is essential to understand that these protocols are not one-size-fits-all. The effectiveness of peptide therapy largely depends on tailoring the cycle length and dosage to the individual's specific health goals, biological responses, and any underlying medical conditions. A cycle refers to the period during which a peptide is actively used, followed by a rest period where the peptide is not administered. This approach helps to maximize the therapeutic benefits while minimizing potential side effects and the risk of developing tolerance to the peptide's effects.

Cycle Length: Typically, peptide cycles can range from as short as a few weeks to as long as several months. For example, a common cycle for bodybuilding peptides like GHRP-6 might last 4-6 weeks, followed by a break of equal length. In contrast, peptides used for anti-aging purposes, such as CJC-1295, may be utilized in longer cycles, often up to 6 months, with a 2-4 week break. It is crucial to start with the lower end of the range for cycle length and gradually adjust based on personal tolerance and response.

Treatment Duration: The overall duration of peptide therapy can vary depending on the individual's goals. Some may seek short-term improvements in muscle recovery and performance, necessitating only a few cycles. Others might aim for long-term benefits in anti-aging and general wellness, which could mean ongoing cycles with periodic evaluations every few months. Regular monitoring through blood tests or physical assessments can help gauge the therapy's effectiveness and guide adjustments in treatment duration.

Rest Periods: Including rest periods between cycles is vital to prevent the body from becoming accustomed to the peptide, which could reduce its effectiveness over time. These breaks also allow the body's natural regulatory mechanisms to reset, potentially reducing the risk of adverse effects. The length of the rest period can be adjusted based on the specific peptide used and the individual's response to therapy.

Dosage Adjustments: Within a cycle, it may be necessary to adjust dosages based on the body's response. Starting with a lower dose and gradually increasing allows for monitoring of both positive effects and potential side effects. If adverse reactions occur, reducing the dosage or discontinuing use may be warranted. Consultation with a healthcare professional can provide guidance on safe dosage adjustments.

Combination Protocols: For those utilizing multiple peptides, it is important to consider how different cycles might overlap or interact. Some peptides can be used concurrently to synergize and enhance each other's effects, while others may need to be staggered to avoid interactions or compounded side effects. Careful planning and consultation with a healthcare provider can help optimize combination protocols for safety and efficacy.

Personalization: Ultimately, the key to successful peptide therapy lies in personalization. Factors such as age, gender, health status, and specific wellness goals should all be considered when planning cycles and treatment durations. Regular evaluation of the therapy's impact, including both subjective and objective measures of improvement, allows for ongoing customization of the protocol to meet the individual's evolving needs.

By adhering to these guidelines, individuals can strategically plan their peptide usage cycles and treatment durations to achieve their health and wellness objectives effectively and safely. It is always recommended to proceed under the guidance of a healthcare professional experienced in peptide therapy to navigate the nuances of cycle planning and to make informed adjustments based on personal progress and feedback.

Planning Usage Cycles

In organizing treatment cycles for peptide therapy, it is crucial to adopt a systematic approach that accounts for the intricate balance between achieving desired outcomes and maintaining the body's natural equilibrium. This necessitates a detailed understanding of each peptide's specific properties, including its half-life, optimal dosage, and the context of its intended use. The half-life of a peptide, which refers to the time it takes for half of the peptide's activity to be reduced by the body, plays a pivotal role in determining the frequency of administration within a cycle. For peptides with a shorter half-life, more frequent dosing may be required to maintain therapeutic levels in the bloodstream, whereas peptides with a longer half-life may necessitate less frequent administration.

Dosage is another critical factor that requires careful consideration. The optimal dosage not only depends on the specific goals of the therapy, whether it be muscle growth, fat loss, or cognitive enhancement, but also on individual factors such as body composition, metabolic rate, and overall health status. It is advisable to begin with the lower end of the recommended dosage range and gradually adjust based on personal response and tolerance. This incremental approach allows for the careful monitoring of effects and side effects, ensuring that the therapy remains both effective and safe.

The **context of use** also significantly influences the planning of peptide cycles. For instance, peptides used in the context of athletic performance or muscle recovery may be cycled in alignment with training periods or competitive seasons, ensuring that the benefits of the peptides are maximized during these critical times. Conversely, peptides utilized for general wellness or anti-aging purposes may follow a more continuous cycle with scheduled breaks to allow the body to reset.

Record-keeping emerges as an indispensable tool in the efficient planning and execution of peptide cycles. By meticulously documenting dosages, administration times, and any observed effects or side effects, individuals can gain insights into their personal response to peptide therapy. This data becomes invaluable when making informed decisions about adjustments to dosages or administration frequencies and when evaluating the overall effectiveness of the treatment.

Consultation with healthcare professionals cannot be overstated in its importance. Collaborating with a practitioner knowledgeable in peptide therapy provides a layer of safety and expertise that enhances the planning process. A healthcare professional can offer personalized advice based on medical history, current health status, and specific therapeutic goals, guiding the structuring of peptide cycles in a manner that is both safe and optimally effective.

The **integration of peptides with other lifestyle factors** such as diet, exercise, and sleep also requires attention during the planning phase. Nutritional support, adequate physical activity, and restorative sleep all play supportive roles in maximizing the benefits of peptide therapy. For example, ensuring the intake of sufficient protein and maintaining hydration can support the body's utilization of peptides, while aligning peptide administration with sleep patterns may enhance certain peptides' effectiveness in promoting restful sleep or recovery.

Adjustments based on feedback and results are a natural and necessary part of managing peptide cycles. The body's response to peptide therapy can evolve over time, necessitating modifications to the treatment protocol. This dynamic process underscores the need for ongoing monitoring and flexibility, allowing for the refinement of peptide cycles to better align with changing health objectives or to address any emerging concerns.

In essence, the successful planning of peptide usage cycles requires a comprehensive, informed approach that considers the specific characteristics of the peptides, the individual's unique physiological and lifestyle factors, and the overarching health and wellness goals. By adhering to a structured yet adaptable framework, individuals can harness the full potential of peptide therapy, achieving significant strides in their health and wellness journeys without compromising safety.

Cycle Strategies for Benefits and Risks

Implementing strategic cycle strategies for peptide use is paramount to harnessing the full spectrum of benefits while concurrently minimizing risks. This necessitates a nuanced understanding of each peptide's unique properties and the biological mechanisms they influence. The optimization of cycle strategies involves not only the meticulous planning of on-cycles and off-cycles but also the consideration of peptide stacking and the potential for synergistic effects. When peptides are used in combination, their collective impact can either amplify the desired outcomes or, if not carefully managed, increase the susceptibility to adverse effects. Therefore, a critical aspect of cycle strategy is the judicious selection of peptides that complement each other's action without exacerbating risk factors.

Synergistic Peptide Stacking: The concept of stacking involves using two or more peptides concurrently to target multiple pathways or achieve a broader range of effects than what might be possible with a single peptide. For instance, combining a peptide known for its muscle-building properties with another that enhances fat loss can potentially offer a comprehensive approach to body composition improvement. However, the key to effective stacking lies in understanding the interaction between peptides. This includes knowledge of their receptor affinities, biological effects, and any overlapping pathways that might lead to unintended consequences. It is advisable to start with the lowest effective dose for each peptide within a stack and closely monitor the body's response to adjust dosages as necessary.

On-Cycle Optimization: During the active phase of peptide use, or the on-cycle, attention must be given to the timing of administration in relation to other peptides in the stack, as well as to lifestyle factors such as diet, exercise, and sleep. Certain peptides exhibit a heightened efficacy when administered at specific times of the day, such as those that influence circadian rhythms or the body's natural hormonal fluctuations. Aligning peptide administration with these biological rhythms can enhance their effectiveness and reduce the risk of desensitization to their effects.

Off-Cycle Planning: The off-cycle, or the period of rest between active cycles, is crucial for allowing the body to reset and reduce the likelihood of developing tolerance or dependence on the effects. The duration of the off-cycle should be proportional to the length and intensity of the on-cycle, with longer and more potent cycles necessitating more extended periods of rest. This phase is also an opportune time to assess the body's response to the treatments, evaluating both the benefits achieved and any adverse effects experienced. Such evaluation can inform future cycle planning, allowing for the fine-tuning of selection, dosages, and combinations.

Risk Mitigation Through Cycle Variation: Varying the substances used across different cycles can help in preventing the body from adapting to their effects, a phenomenon that can lead to diminished results over time. Introducing new compounds or altering the combination of substances between cycles can stimulate the body in new ways, potentially leading to continued progress towards health and wellness goals. This strategy also reduces the risk of long-term side effects associated with prolonged use of a single compound or combination.

Monitoring and Adjustments: Continuous monitoring throughout the peptide therapy process is indispensable. This includes tracking progress towards health objectives, noting any side effects, and making necessary adjustments to the cycle strategy based on these observations. Regular consultations with healthcare professionals experienced in peptide therapy can provide valuable insights and guidance, ensuring that the cycle strategy remains aligned with the individual's evolving health and wellness needs.

Personalization of Cycle Strategies: Ultimately, the most effective cycle strategy is one that is highly personalized. This requires taking into account not only the individual's specific health and wellness objectives but also their unique physiological responses to peptide therapy. Factors such as age, gender, baseline health status, and personal tolerance levels must be considered when devising a cycle strategy. Tailoring the approach to the individual ensures that the benefits of peptide therapy are maximized while the risks are kept to a minimum.

Chapter 8: Combined Protocols and Synergies

Synergistic Peptides

Understanding the nuances of synergistic peptides involves not just identifying which substances can be used together but also grasping the intricacies of their interactions to optimize health outcomes. The concept of synergy in therapy is predicated on the idea that certain agents, when combined, produce a compounded effect that is greater than the sum of their individual effects. This synergistic relationship can be leveraged to amplify benefits across various domains, including but not limited to muscle recovery, anti-aging, and cognitive function. To effectively harness the power of synergistic compounds, one must consider several key factors, including their mechanism of action, their receptor targets, and the physiological pathways they influence.

For instance, the combination of CJC-1295 and Ipamorelin is a prime example of a synergistic duo that is frequently utilized to enhance growth hormone secretion CJC-1295 acts as a GHRH (Growth Hormone-Releasing Hormone) analog, extending the half-life of GHRH and its effects, while Ipamorelin, a GHRP (Growth Hormone Releasing Agent), mimics ghrelin and stimulates the release of growth hormone from the pituitary gland. When used together, these substances create a powerful stimulus for growth hormone release, far exceeding the response elicited by either agent alone. This synergistic effect not only maximizes the potential for muscle growth and fat loss but also reduces the required dosages of each substance, potentially mitigating the risk of side effects.

Another compelling synergy is observed between compounds that target skin health. The combination of Matrixyl (palmitoyl pentapeptide-4) and Copper agents offers a robust approach to anti-aging skincare. Matrixyl promotes the production of collagen and elastin, essential proteins that provide skin its firmness and elasticity, while Copper agents not only stimulate skin regeneration and healing but also enhance the anti-inflammatory and antioxidant responses. This combination addresses skin aging from multiple angles, improving skin texture, reducing the appearance of wrinkles, and promoting a youthful complexion.

In the realm of cognitive enhancement, the strategic use of peptides like **Semax** and **Selank** demonstrates the potential for synergistic effects on mental clarity and anxiety reduction. Semax acts to increase brain-derived neurotrophic factor (BDNF), thereby supporting neuronal growth and cognitive function, while Selank, an anxiolytic peptide, reduces anxiety without sedative effects. Together, these peptides can improve cognitive performance while also providing a calming effect, making them a potent combination for those seeking mental enhancement alongside stress relief.

To effectively implement synergistic peptide protocols, it is essential to adopt a methodical approach, beginning with a thorough understanding of each peptide's profile, including its pharmacokinetics, pharmacodynamics, and any known interactions. This knowledge base enables the identification of potentially synergistic pairs or groups of peptides. Following this, a carefully structured regimen should be designed, taking into account the optimal dosing schedules, routes of administration, and monitoring protocols to assess efficacy and safety. Regular evaluation and adjustment of the regimen are crucial, as individual responses to peptide therapy can vary widely.

Moreover, it is important to recognize that while synergistic peptide combinations hold great promise for enhancing health and wellness outcomes, they also require a heightened level of vigilance regarding potential interactions and side effects. Consulting with healthcare professionals experienced in peptide therapy is strongly recommended to navigate the complexities of synergistic peptide use safely and effectively.

In conclusion, the strategic combination of these bioactive compounds to exploit synergistic effects represents a frontier in personalized health and wellness optimization. By carefully selecting and combining these substances with complementary actions, individuals can achieve enhanced results, whether the goal is improved physical performance, accelerated recovery, anti-aging benefits, or cognitive enhancements. However, the successful application of synergistic compounds demands a deep understanding of their science, precise protocol design, and ongoing monitoring to ensure safety and efficacy.

Effective Health Combinations

Building on the foundational understanding of peptides and their individual benefits, it becomes evident that their strategic combination can amplify outcomes, particularly in areas critical to our health-conscious audience: muscle recovery and brain health. The synergy between specific peptides can create a more holistic approach to wellness, addressing not just isolated issues but promoting overall well-being.

For muscle recovery, the combination of **MGF (Mechano Growth Factor)** and **IGF-1 (Insulin-like Growth Factor-1)** offers a potent duo. MGF initiates muscle satellite cell activation, essential for muscle repair and growth following exercise, while IGF-1 promotes both muscle growth and repair by stimulating amino acid uptake and protein synthesis in muscle cells. This synergistic effect not only accelerates recovery time but also enhances the muscle growth response to resistance training, making it an invaluable protocol for individuals aiming to improve physical performance and muscle health.

Turning to brain health, the pairing of **CJC-1295** and **Ipamorelin** is noteworthy. CJC-1295, a GHRH (Growth Hormone-Releasing Hormone) analog, along with Ipamorelin, a selective GHRP (Growth Hormone Releasing Peptide), work together to increase levels of growth hormone in the body. Elevated growth hormone levels have been associated with improved cognitive function, memory retention, and neuroprotection, offering a protective buffer against age-related cognitive decline. This combination not only supports the structural health of the brain but also enhances cognitive resilience, making it a strategic choice for those seeking to preserve mental clarity and function.

Furthermore, the integration of peptides targeting both muscle and brain health into a single regimen can have compounding benefits. For example, improved muscle health contributes to overall metabolic efficiency, which in turn supports cognitive function by ensuring a steady supply of nutrients and oxygen to the brain. Conversely, a healthy, well-functioning brain supports the motivation and cognitive functions necessary for maintaining an active lifestyle and adhering to a consistent exercise regimen, thereby promoting muscle health.

When considering the administration of these peptides, it is crucial to adhere to recommended dosages and administration methods. Subcutaneous injections are commonly used for both MGF and IGF-1 due to their efficient absorption and ability to maintain potency. Similarly, CJC-1295 and Ipamorelin are typically administered via subcutaneous injections to ensure optimal bioavailability and effectiveness. The timing of these administrations should be carefully planned to align with the body's natural rhythms and the individual's lifestyle for maximum benefit. For instance, administering certain peptides before bedtime can leverage the body's natural growth hormone release cycle, enhancing the restorative processes that occur during sleep.

It is also essential to monitor the body's response to these peptide combinations. Adjustments to dosages or administration schedules may be necessary to optimize outcomes and minimize any potential side effects. Regular consultation with a healthcare professional experienced in peptide therapy can provide personalized guidance, ensuring that the chosen peptide regimen is both safe and effective for the individual's unique health profile.

In conclusion, the strategic combination of peptides for muscle recovery and brain health exemplifies the potential of synergistic peptide protocols to address complex health objectives. By understanding the mechanisms through which these peptides operate and carefully planning their combined use, individuals can harness the full potential of peptides to support their health and wellness goals.

Avoiding Negative Interactions

To **avoid negative interactions** when using peptides, it is paramount to understand the biochemical individuality of each person. This means that the way one's body responds to a certain regimen can significantly differ from another's. As such, a meticulous approach to combining these compounds is crucial. Firstly, it is essential to recognize that these substances, while beneficial on their own, can lead to adverse effects if improperly combined. For instance, certain compounds may compete for the same receptors in the body, thereby diminishing their overall effectiveness or leading to unexpected side effects. To circumvent this, one should adhere to the principle of starting with a singular compound, observing the body's response, and only then, cautiously adding another into the regimen.

Moreover, the timing of administration plays a critical role in avoiding negative interactions. For example, substances that stimulate the release of growth hormone should not be administered close in time to those that might inhibit or modulate this release in an undesired manner. This careful scheduling ensures that the compounds do not counteract each other's effects but rather work in synergy to enhance overall health and wellness.

Another consideration is the route of administration. Different substances have varying degrees of bioavailability depending on whether they are injected, taken orally, or applied topically. This can influence not only their efficacy but also their potential to interact negatively with other compounds or medications. For instance, subcutaneous injections provide a direct route into the bloodstream, which might lead to more pronounced interactions with other substances compared to topical applications that primarily affect localized areas.

It is also advisable to consult with a healthcare professional experienced in this therapy before starting any new regimen, especially when planning to combine multiple compounds. This consultation should include a thorough review of the individual's current medications, supplements, and overall health status to identify any potential contraindications or risks.

Furthermore, monitoring the body's response to therapy is essential. This involves paying close attention to any signs of adverse reactions or side effects and adjusting the regimen accordingly. Regular blood work and other diagnostic tests can provide objective data on how the body is responding and help in making informed decisions about whether to continue, adjust, or discontinue certain substances.

Lastly, it is crucial to source these compounds from reputable suppliers to ensure their purity and potency. Impurities and variations in quality can not only diminish their effectiveness but also increase the risk of negative interactions. Therefore, verifying the source and ensuring the quality of these substances is a step that cannot be overlooked.

By taking these measures, individuals can maximize the benefits of therapy while minimizing the risk of negative interactions. This careful, informed approach to combining these powerful molecules supports the goal of enhancing health and wellness through their strategic use.

Choosing Safe and Compatible Peptides

Choosing safe and compatible compounds requires a nuanced understanding of each substance's specific mechanism of action and its potential interactions within the body's complex biochemical environment. This understanding is foundational to formulating a regimen that not only achieves desired health outcomes but also maintains the highest safety standards. When selecting these compounds, one must consider several critical factors to ensure compatibility and safety, emphasizing the importance of a methodical and evidence-based approach.

The first step in selecting safe and compatible substances involves a thorough review of the scientific literature to identify those with a well-documented track record of safety and efficacy. This review should focus on clinical studies, peer-reviewed articles, and case reports that provide comprehensive insights into the compounds' pharmacokinetics—how they are absorbed, distributed, metabolized, and excreted by the body—and pharmacodynamics, which detail the biochemical and physiological effects of the substances, including their mechanisms of action. Such a meticulous review helps in understanding both the potential benefits and risks associated with each compound, laying the groundwork for informed decision-making.

Equally important is the assessment of the source and purity of the substances. Given the unregulated nature of many suppliers, especially those operating online, it is imperative to procure these compounds from reputable sources that provide third-party testing results confirming the purity and potency of their products. This step cannot be overstated, as impurities and variations in quality can significantly impact both their effectiveness and safety profile.

Once the substances of interest are identified and sourced, the next step is to consider the individual's unique health status, including any underlying medical conditions, current medications, and overall health objectives. This personalized health assessment is crucial because it helps identify any potential contraindications or risks that could arise from use. For instance, individuals with a history of certain cancers may need to avoid compounds that stimulate growth hormone production due to the theoretical risk of accelerating tumor growth. Similarly, substances that influence insulin sensitivity must be used with caution in individuals with diabetes or metabolic syndrome.

The timing and route of administration also play a pivotal role in ensuring compatibility and safety. Certain compounds may exhibit optimal efficacy when administered at specific times of the day, such as those that align with the body's natural circadian rhythms of hormone production. Additionally, the route of administration—whether subcutaneous injection, oral, topical, or nasal spray—affects the bioavailability and, consequently, the interaction with other substances or medications. Understanding these nuances is essential for developing a regimen that maximizes therapeutic benefits while minimizing risks.

Moreover, starting with a conservative approach by introducing one compound at a time allows for the careful observation of the body's response. This stepwise approach facilitates the identification of any adverse effects or intolerances early in the regimen, thereby enabling timely adjustments before introducing additional substances. It also provides valuable insights into the efficacy of each compound, helping to tailor the regimen more precisely to the individual's needs.

Finally, ongoing monitoring and adjustment of the regimen are imperative. Regular follow-ups with a healthcare professional experienced in this therapy can help assess the regimen's effectiveness and safety, making necessary adjustments based on the individual's evolving health status and objectives. This dynamic approach ensures that the regimen remains both safe and aligned with the individual's wellness journey.

In essence, selecting safe and compatible substances is a multifaceted process that demands a comprehensive understanding of their properties, a careful consideration of the individual's health status, and a commitment to ongoing monitoring and adjustment. By adhering to these principles, individuals can harness the potential of these compounds to support their health and wellness goals while minimizing the risk of negative interactions and maximizing safety.

Chapter 9: Monitoring and Optimizing Results

Measuring Protocol Effectiveness

To accurately measure the effectiveness of peptide protocols, it's imperative to implement a comprehensive monitoring strategy that encompasses both subjective assessments and objective measurements. This multifaceted approach ensures that individuals can track their progress effectively, making informed decisions about their peptide regimen's continuation, adjustment, or cessation.

Subjective Assessments involve personal observations and reflections on one's physical and mental state. Individuals should maintain a detailed journal documenting their baseline health status before initiating peptide therapy, including notes on specific concerns such as energy levels, cognitive function, skin elasticity, and overall well-being. As the regimen progresses, regular entries highlighting changes, improvements, or any adverse reactions provide invaluable insights into the peptides' impact. This personalized record serves as a crucial tool for evaluating subjective improvements in areas like mental clarity, mood, sleep quality, and perceived muscle recovery times.

Objective Measurements, on the other hand, offer quantifiable data that can corroborate subjective assessments. These include:

Comprehensive blood panels can reveal changes in biomarkers related to inflammation, hormone levels, and organ function, which may indicate the physiological effects of these compounds. Monitoring changes in specific markers, such as growth hormone levels and insulin-like growth factor 1 (IGF-1), can be particularly telling for substances aimed at enhancing muscle growth or anti-aging effects.

Tools like DEXA scans provide detailed insights into body composition, including fat mass, muscle mass, and bone density. Such analysis allows for the objective evaluation of these compounds' effects on muscle growth, fat loss, and overall body composition, offering a clear picture of physical changes over time.

For substances targeting brain health, cognitive function tests available online or through healthcare providers can quantify changes in memory, attention, and other cognitive abilities. Regular testing before and during therapy can help gauge improvements in cognitive functions, supporting subjective observations of enhanced mental clarity and focus.

In the context of compounds for skin health and anti-aging, professional skin analysis techniques, such as high-resolution imaging and elasticity measurements, can objectively assess changes in skin texture, elasticity, and appearance. These measurements complement subjective observations of skin improvements, providing a basis for evaluating the effectiveness of skincare substances.

For individuals focused on muscle recovery and physical performance, tracking workout performance, recovery times, and strength gains offers objective evidence of benefits. Technologies ranging from wearable fitness trackers to professional assessments can quantify improvements in physical capabilities, aligning with subjective perceptions of enhanced performance and reduced recovery times.

Incorporating both subjective assessments and objective measurements into a comprehensive monitoring strategy empowers individuals to evaluate the effectiveness of their protocols accurately. This dual approach facilitates informed decisions regarding the adjustment of dosages, the introduction of new substances, or the cessation of specific compounds, based on a holistic view of their impacts on health and wellness. Regular consultation with healthcare professionals experienced in this therapy is also crucial, providing expert guidance and interpretation of both subjective and objective data to optimize regimen outcomes.

Evaluating Peptide Impact Techniques

Evaluating the impact of peptide therapy on one's health requires a detailed and methodical approach, leveraging both the subjective assessments and objective measurements discussed previously. To further refine this evaluation process, it's essential to understand the importance of **longitudinal tracking**. This involves consistent monitoring over extended periods, which is particularly crucial for capturing the gradual effects of peptides on health markers that may not exhibit immediate changes. For instance, the benefits of peptides on skin elasticity and texture might take several weeks or months to become noticeable. Therefore, establishing a baseline before starting peptide therapy and conducting follow-up assessments at regular intervals, such as monthly or quarterly, can provide a clearer picture of the peptides' effectiveness over time.

Another critical technique in evaluating peptide impact involves **comparative analysis**. This method compares health markers before and after the initiation of peptide therapy, offering a direct way to assess changes. For example, comparing pre-therapy and post-therapy DEXA scan results can quantitatively demonstrate changes in body composition, such as increases in muscle mass or reductions in fat mass. Similarly, cognitive function tests conducted before and during peptide therapy can reveal improvements in memory, attention, and other cognitive abilities, providing concrete evidence of the peptides' benefits on brain health.

Peer comparison can also serve as a valuable tool in evaluating peptide impact. This involves comparing one's progress and results with those of others following similar peptide protocols. While individual responses to peptide therapy can vary widely, understanding the range of possible outcomes and where one's results fall within that spectrum can offer additional insights into the effectiveness of their regimen. Engaging with online forums, support groups, or consulting with a healthcare professional who has experience with multiple patients on peptide therapy can facilitate this type of comparison.

Incorporating **technology and apps** designed for health tracking can significantly enhance the precision and ease of monitoring the effects of peptide therapy. Numerous apps and devices are available that track various health metrics, such as sleep quality, heart rate variability, physical activity levels, and more. These tools can automate the process of data collection, providing a wealth of information that can be analyzed to detect subtle changes and trends over time. For instance, a consistent improvement in sleep quality or a decrease in resting heart rate over several months of peptide therapy can be strong indicators of the regimen's positive impact on one's health.

Feedback from healthcare professionals plays a pivotal role in evaluating the impact of peptides. Regular consultations with doctors or specialists who are knowledgeable about peptide therapy can provide expert insights into the observed changes and help contextualize them within the broader scope of one's health and wellness goals. These professionals can also recommend adjustments to the peptide regimen based on the latest research and clinical guidelines, ensuring that the therapy remains aligned with the individual's evolving health needs.

By employing these practical techniques, individuals can conduct a comprehensive evaluation of the impact of peptide therapy on their health. This rigorous approach not only helps in assessing the effectiveness of the current regimen but also informs future decisions regarding the continuation, adjustment, or expansion of peptide use. Through careful monitoring and evaluation, individuals can maximize the benefits of peptide therapy, ensuring that their health and wellness objectives are met with the highest degree of safety and efficacy.

Personalized Protocol Adjustments

Personalized adjustments to peptide protocols are essential for optimizing health outcomes and ensuring safety. The key to successful personalization lies in the careful observation of one's responses to peptide therapy and the willingness to make adjustments based on these observations. This process involves several steps, starting with the initial assessment and continuing through regular monitoring and fine-tuning of the regimen.

The initial assessment should include a comprehensive evaluation of one's health status, including any pre-existing conditions, allergies, and sensitivities. This assessment serves as the baseline against which the effects of the peptide regimen can be measured. It is also crucial to consider lifestyle factors such as diet, exercise habits, and stress levels, as these can influence the body's response to peptides.

Once the peptide therapy has begun, it is important to monitor both the subjective and objective markers of health. Subjective markers include perceived changes in energy levels, mood, mental clarity, and overall well-being. Objective markers, on the other hand, can be monitored through regular blood tests, body composition analysis, and other diagnostic tools. These markers provide quantifiable data on the effects of the peptide regimen on various aspects of health, including hormone levels, inflammation, muscle mass, and fat distribution.

Based on the data gathered from these assessments, adjustments to the regimen may be necessary. For example, if blood tests show an unexpected increase in certain hormones, the dosage of the compounds may need to be adjusted. Similarly, if subjective assessments indicate an improvement in energy levels but no change in body composition, the combination of substances being used may need to be reevaluated.

The timing and route of administration may also require adjustments. Some compounds may be more effective when taken at certain times of the day or in conjunction with specific activities, such as exercise. Additionally, the method of administration—whether subcutaneous injection, oral, topical, or nasal spray—can affect the bioavailability and efficacy of the substances. Experimenting with different timing and administration methods, under the guidance of a healthcare professional, can help optimize the benefits of therapy.

Another important aspect of personalizing protocols is the consideration of synergistic effects. Some substances work better in combination with others, enhancing their effects on muscle growth, fat loss, cognitive function, or skin health. Careful selection and combination of compounds, based on one's specific health goals and the observed effects of the regimen, can lead to more effective and personalized outcomes.

It is also essential to consider the duration of therapy. Some compounds may yield the desired effects only after extended use, while others may produce quick results that do not last without ongoing administration. Adjusting the duration of therapy, as well as planning for periodic reassessment and possible cycle adjustments, is crucial for maintaining and optimizing the benefits of treatment.

Finally, personalizing peptide protocols requires a commitment to ongoing education and consultation with healthcare professionals. As new research emerges and our understanding of peptides evolves, staying informed and seeking expert guidance can help ensure that one's peptide regimen remains safe, effective, and aligned with the latest scientific findings.

Personalizing peptide protocols is a dynamic process that requires careful monitoring, willingness to make adjustments, and ongoing consultation with healthcare professionals. By taking a proactive and informed approach to peptide therapy, individuals can optimize their health outcomes and achieve their wellness goals.

Modifying Protocols for Personal Results

To effectively modify peptide protocols based on personal results, it is imperative to meticulously document and analyze individual responses to peptide therapy over time. This process involves a comprehensive approach, beginning with the initial assessment of personal health status and specific wellness objectives. Establishing a baseline through laboratory tests and physical assessments prior to initiating peptide therapy allows for a clear comparison of pre- and post-therapy metrics, enabling individuals to gauge the effectiveness of the peptides on their health and wellness journey.

Monitoring one's progress requires consistent tracking of various parameters, including but not limited to, physical performance metrics, body composition analyses, subjective wellness surveys, and regular blood work. These tools collectively provide a holistic view of how the body is responding to peptide therapy. For instance, improvements in muscle recovery rates, changes in body fat percentage, enhanced cognitive function, and alterations in sleep quality can all indicate the efficacy of the peptide regimen. It is crucial to note any adverse reactions or side effects, as these may necessitate immediate adjustments to the protocol.

Personal feedback plays a pivotal role in optimizing peptide therapy. This includes not only tangible changes measured through tests and assessments but also subjective experiences such as energy levels, mood improvements, and overall sense of well-being. Such feedback, when discussed with a healthcare professional knowledgeable in peptide therapy, can guide the fine-tuning of peptide types, dosages, and administration schedules.

Adjustments to the peptide protocol should be made cautiously and incrementally to isolate variables and understand the impact of each change. For example, if increasing the dosage of a specific peptide does not yield the expected improvement or leads to side effects, a reduction in dosage or a switch to an alternative peptide may be considered. Similarly, the timing of administration may be adjusted based on personal schedules and the body's response to optimize absorption and effectiveness.

The integration of peptides with other lifestyle modifications such as diet, exercise, and stress management techniques should also be revisited. Peptides do not work in isolation; their effects can be significantly enhanced or hindered by other aspects of an individual's lifestyle. Nutritional adjustments may be required to support the body's needs during peptide therapy, while changes in exercise routines can further enhance muscle recovery and growth facilitated by peptides. Stress reduction techniques and sleep hygiene practices may amplify the cognitive and anti-aging benefits of peptide therapy.

In conclusion, the key to modifying peptide protocols based on personal results lies in a disciplined, data-driven approach to tracking one's health and wellness journey. By combining objective data with subjective feedback and working closely with healthcare professionals, individuals can tailor their peptide therapy to achieve optimal results tailored to their unique physiological makeup and wellness goals. Continuous learning and adaptation, guided by personal experiences and scientific evidence, are essential for maximizing the benefits of peptide therapy in pursuit of lasting health and wellness.

Chapter 10: Potential Risks and Side Effects

Common Side Effects

Monitoring for common side effects is an essential aspect of safely incorporating peptides into your wellness regimen. **Side effects** can vary widely depending on the type of peptide, dosage, and individual response. Some of the most frequently encountered side effects include **irritation at the injection site, fatigue, headaches,** and **dry mouth**. Less commonly, individuals may experience **nausea, dizziness**, and changes in appetite. It's crucial to understand that while peptides offer promising benefits for health and wellness, their impact on the body must be carefully managed.

Irritation at the injection site is often reported by those new to peptide therapy, especially with subcutaneous injections. This irritation can manifest as redness, swelling, or discomfort. To mitigate these effects, rotate injection sites regularly and consider applying a cold compress to the area post-injection. If irritation persists, consult with a healthcare professional to ensure proper technique and to discuss the possibility of allergic reactions.

Fatigue and **headaches** are also common and can be attributed to the body's adjustment to the peptides or the modulation of certain biological processes. Staying well-hydrated and maintaining a balanced diet can help manage these symptoms. If they become bothersome or persistent, adjusting the dosage or timing of administration might be necessary under the guidance of a healthcare provider.

Dry mouth, while less common, can be an indication of the body's adaptation to certain peptides that influence hydration and salivation. Increasing water intake and using saliva-promoting products can alleviate this discomfort.

For those experiencing **nausea** or **dizziness**, these symptoms can often be minimized by adjusting the dose or the timing of peptide administration. Taking peptides with a small amount of food or before bedtime may also help reduce these side effects.

Changes in **appetite** can occur, particularly with peptides known to modulate ghrelin and leptin, the hormones responsible for hunger signals. Monitoring dietary intake and maintaining a schedule for meals can help manage unexpected shifts in appetite.

It is essential to start with lower doses of peptides and gradually increase as needed, allowing the body to adjust and minimizing the risk of side effects. Regular consultation with a healthcare professional experienced in peptide therapy is paramount to safely navigating these treatments. They can provide personalized advice and adjustments based on your body's responses and health goals.

Remember, the key to minimizing side effects lies in careful monitoring, open communication with healthcare providers, and a willingness to adjust protocols as needed. Keeping a detailed journal of your peptide use, including doses, timing, and any side effects experienced, can be invaluable in optimizing your approach to peptide therapy and achieving your wellness objectives.

Signs and Symptoms to Monitor

Continuing from the discussion on common side effects associated with peptide therapy, it is essential to delve deeper into the specific signs and symptoms that warrant close monitoring. These indicators not only help in identifying adverse reactions early but also play a critical role in ensuring that peptide use remains within the safe and beneficial spectrum of health and wellness practices. Given the diversity of peptides and their myriad applications, the range of potential side effects can be broad, necessitating a vigilant approach to self-observation and reporting.

One of the initial signs to be mindful of includes any form of hypersensitivity reaction. This could manifest as hives, excessive swelling beyond the injection site, or an unexpected rash. Such reactions could indicate an allergic response to the peptide itself or one of the components used in its formulation. Immediate cessation of the peptide in question and consultation with a healthcare provider are crucial steps in addressing this concern.

Another critical area of monitoring involves cardiovascular symptoms. For some individuals, peptides can influence blood pressure and heart rate. Symptoms such as palpitations, an unusual spike or drop in blood pressure, or an unexplained feeling of lightheadedness should prompt a reevaluation of the peptide regimen. These symptoms, while rare, underscore the importance of starting with lower doses and gradually titrating up, under medical supervision, to gauge the body's response.

Gastrointestinal disturbances can also occur, particularly with peptides that influence digestive processes or appetite regulation. Symptoms may include abdominal discomfort, significant changes in bowel habits, or unexplained nausea that persists beyond the initial adjustment period. While these symptoms are often temporary, they should not be dismissed, especially if they persist or worsen over time.

Mood changes and sleep disturbances represent another category of symptoms to monitor closely. Peptides, by virtue of their action on various hormones and neurotransmitters, can impact mood and emotional well-being. Feelings of anxiety, unusual irritability, or disturbances in sleep patterns that coincide with peptide therapy should be carefully evaluated. Adjustments to dosage or the discontinuation of certain peptides may be warranted based on these observations.

It's also pertinent to monitor for symptoms that could suggest endocrine disruption, such as unexplained weight gain or loss, changes in hair texture or growth, and menstrual irregularities in women. Given that peptides can interact with hormone levels, keeping a detailed record of such changes can provide valuable insights into how the body is responding to peptide therapy.

Monitoring renal and hepatic function through regular blood work is advisable, especially for those on long-term peptide therapy. While peptides are generally well-tolerated, ensuring that the liver and kidneys are not under undue stress is important. Any significant changes in laboratory markers related to these organs should be discussed with a healthcare provider to determine the appropriateness of continuing peptide therapy.

Lastly, cognitive changes, though uncommon, should not be overlooked. This includes significant changes in memory, attention, or cognitive processing that are noted after commencing peptide therapy. While some peptides are designed to enhance cognitive function, any adverse changes should prompt a reevaluation of the peptide regimen being used.

In all instances, the key to safely navigating the use of peptides lies in open and ongoing communication with a healthcare professional experienced in peptide therapy. This partnership enables the tailoring of peptide regimens to individual needs and conditions, enhancing the potential benefits while minimizing risks. It also emphasizes the importance of personal responsibility in monitoring one's health and being proactive in reporting any adverse symptoms. Through such vigilant practices, the integration of peptides into a health and wellness routine can be optimized, aligning with the goal of achieving improved health outcomes without compromising safety.

Safety Warnings and Precautions

Embarking on a peptide therapy regimen without the guidance of a healthcare professional can lead to unforeseen complications and diminish the potential benefits of these powerful molecules. It is paramount to consult with a doctor who is knowledgeable in peptide therapy before starting any new peptide regimen. This consultation is not merely a formality but a crucial step in ensuring that your peptide usage aligns with your specific health conditions and wellness goals. A healthcare provider can offer invaluable insights into the most appropriate peptides for your needs, the optimal dosages, and the safest administration routes. They can also identify any potential interactions with existing medications or underlying health conditions that could contraindicate peptide use.

Another critical aspect of therapy safety involves sourcing these compounds from reputable suppliers. The market for such substances is vast, and not all suppliers adhere to the stringent quality control standards necessary for products intended for therapeutic use. Purchasing from unreliable sources can expose you to products of dubious purity and potency, increasing the risk of adverse reactions and diminishing the potential health benefits. Always verify the credibility of the supplier, looking for evidence of third-party testing and quality certification. This due diligence ensures that the substances you use are safe, effective, and free from contaminants that could harm your health.

Proper storage and handling of these compounds are also essential safety considerations. They are complex biological molecules that can degrade if not stored correctly, potentially leading to reduced efficacy or increased risk of side effects. Follow the manufacturer's instructions for storage, paying close attention to temperature requirements and expiration dates. Mishandling these substances can compromise their integrity, so it's crucial to adhere to best practices for preparation and administration, whether that involves reconstituting lyophilized forms or ensuring sterility in the injection process.

Adhering to recommended dosages and administration guidelines is another cornerstone of safe use. The allure of accelerated results can tempt some individuals to exceed the recommended dosages or to combine multiple substances without professional guidance. Such practices can lead to adverse effects and diminish the long-term sustainability of therapy. It's essential to start with the lowest effective dose and adjust gradually under medical supervision, allowing your body to acclimate to the compounds and minimizing the risk of side effects.

Monitoring your body's response to therapy is an ongoing responsibility. Even with the most careful planning and professional guidance, individual reactions can vary. Regular check-ins with your healthcare provider, coupled with self-monitoring for any adverse reactions or unexpected changes in your health, are crucial. This vigilance enables timely adjustments to your regimen, enhancing safety and efficacy.

In summary, the responsible use of these substances for health and wellness hinges on professional medical guidance, careful sourcing, proper handling, adherence to recommended practices, and vigilant monitoring. By prioritizing these safety precautions, individuals can explore the benefits of therapy with confidence, making informed decisions that support their health and wellness goals.

Importance of Consulting a Doctor

Consulting a healthcare professional before embarking on a peptide therapy regimen is a critical step that cannot be overstated. The reasons for this are multifaceted and rooted in ensuring the safety, efficacy, and appropriateness of peptide use for an individual's specific health objectives and conditions. A doctor, particularly one with expertise in peptide therapy, can provide a level of insight and guidance that is indispensable for navigating the complex landscape of peptide use. This professional oversight is crucial for several reasons, including the assessment of medical history, the identification of potential interactions with existing medications, and the customization of peptide protocols to align with individual health goals and physiological responses.

The initial consultation with a healthcare provider serves as an opportunity to thoroughly assess one's medical history and current health status. This comprehensive evaluation is essential for identifying any underlying conditions that may contraindicate the use of certain peptides. For instance, individuals with a history of certain cancers, insulin resistance, or autoimmune disorders may require careful consideration and specialized guidance when it comes to peptide therapy. A healthcare professional can identify these nuances and advise accordingly, ensuring that peptide use does not exacerbate existing conditions or introduce new health risks.

Moreover, the potential for interactions between peptides and existing medications is a concern that necessitates medical oversight. Peptides, due to their biological nature, can influence a wide range of physiological processes and potentially interact with medications, either diminishing their efficacy or exacerbating their effects. A healthcare provider can review one's current medication regimen and make informed recommendations to avoid adverse interactions, adjusting medication dosages or advising on the timing of peptide administration as needed.

Customization of peptide protocols is another area where the expertise of a healthcare professional is invaluable. Given the diversity of peptides available, each with its own specific mechanisms of action and effects, a one-size-fits-all approach is not viable. A doctor can tailor peptide protocols, selecting peptides that align with the individual's health goals, whether it be enhancing muscle recovery, supporting weight management, improving cognitive function, or addressing skin health and anti-aging concerns. This personalized approach not only optimizes the potential benefits of peptide therapy but also minimizes the risk of side effects.

Furthermore, a healthcare professional can guide the determination of appropriate dosages and administration routes for peptides, which can significantly impact their efficacy and safety. The optimal dosage of peptides can vary widely among individuals, influenced by factors such as age, weight, health status, and specific health objectives. A doctor can recommend starting dosages and adjust them based on the individual's response to therapy, ensuring that the therapeutic benefits are maximized while minimizing any adverse effects.

The importance of ongoing monitoring and adjustment of peptide protocols underlines the necessity of regular consultations with a healthcare provider. As individuals progress with peptide therapy, their responses to treatment should be closely monitored, and adjustments to the protocol may be necessary. A healthcare professional can conduct or recommend regular health assessments, including laboratory tests, to monitor the effects of peptide therapy on various health markers and make timely adjustments to the treatment plan based on these results.

In essence, the role of a healthcare professional in guiding peptide therapy cannot be understated. Their expertise ensures that individuals embark on peptide therapy with a clear understanding of the potential benefits and risks, armed with personalized protocols that offer the best chance of achieving their health and wellness goals while safeguarding their health. This professional guidance is a cornerstone of responsible peptide use, ensuring that individuals navigate their health journey with informed, evidence-based strategies tailored to their unique needs and conditions.

Contraindications in Peptide Use

When considering peptide therapy, it's imperative to be aware of specific medical conditions that may necessitate caution or outright avoidance of certain peptides. These contraindications are not merely guidelines but are critical for ensuring safety and preventing adverse effects that could arise from peptide use in vulnerable populations. First and foremost, individuals with a history of cancer should exercise extreme caution. Peptides, particularly those that stimulate growth hormone production, could potentially accelerate the growth of existing tumors or cancerous cells due to their proliferative effects. Therefore, anyone with a current diagnosis or history of cancer must consult with a healthcare provider before initiating peptide therapy.

Autoimmune diseases present another area of concern. Since peptides can modulate the immune system, those with conditions like rheumatoid arthritis, lupus, or multiple sclerosis should approach peptide use with caution. The risk lies in the possibility of peptides either exacerbating the autoimmune response or, conversely, suppressing the immune system too much. A tailored, closely monitored approach is essential for individuals with autoimmune diseases, ensuring that peptide therapy does not interfere with their condition's management.

Individuals with a history of heart disease or those at risk for cardiovascular issues should also be cautious. Certain peptides can influence blood pressure, heart rate, and vascular health, potentially posing risks for those with pre-existing heart conditions. It's crucial for individuals with cardiovascular concerns to undergo a thorough medical evaluation before considering peptides, focusing on those that might not exacerbate their condition.

For those with kidney or liver diseases, peptides that are metabolized or excreted via these organs require careful consideration. The added metabolic load could impair organ function further, highlighting the need for medical guidance and possibly adjusted dosages or specific peptide selections to mitigate risks.

Lastly, pregnant or breastfeeding women are advised against using peptides due to the lack of research on peptide therapy's effects during pregnancy and lactation. The potential impact on fetal development or infant health through breast milk remains largely unknown, warranting a cautious approach until more information becomes available.

In summary, while peptides offer promising benefits for health and wellness, their use must be tailored to individual health profiles, especially in the presence of contraindications. Consulting with healthcare professionals, selecting peptides with an understanding of their effects, and closely monitoring therapy's impact are vital steps for those with underlying medical conditions. By adopting a cautious and informed approach, individuals can maximize the benefits of peptide therapy while minimizing potential risks.

Health Conditions Requiring Attention

Individuals with endocrine disorders, such as diabetes or thyroid conditions, warrant special consideration when exploring hormone therapy. The intricate balance of hormones within the endocrine system can be sensitive to external influences, including substances known to modulate hormonal pathways. For those managing diabetes, therapies that influence insulin sensitivity or glucose metabolism must be approached with caution. The potential for these treatments to either enhance or diminish insulin sensitivity underscores the necessity for rigorous blood glucose monitoring and possibly the adjustment of diabetes management plans under medical supervision. Similarly, individuals with thyroid disorders should be aware that certain therapies might interact with thyroid function, either by amplifying or suppressing the production of thyroid hormones. Given the thyroid's central role in regulating metabolism, energy levels, and overall hormonal balance, any regimen should be carefully considered and monitored in collaboration with healthcare providers specializing in endocrine health.

Respiratory conditions, including asthma and chronic obstructive pulmonary disease (COPD), also present a scenario where the use of such therapies must be judiciously evaluated. Treatments that have an influence on inflammation and immune response could have implications for respiratory health, potentially offering benefits in reducing airway inflammation but also carrying the risk of unintended effects on immune function that could exacerbate respiratory symptoms. Patients with chronic respiratory conditions should engage in a detailed dialogue with their healthcare team to weigh the potential risks and benefits of these therapies, ensuring any chosen options align with their overall treatment strategy and do not compromise respiratory function.

Furthermore, individuals with psychiatric or neurological conditions, such as depression, anxiety, or neurodegenerative diseases like Alzheimer's, may find that certain therapies influence their symptoms in unpredictable ways. Treatments that cross the blood-brain barrier and interact with neurotransmitters or neuroreceptors can potentially affect mood, cognitive function, and neurological health. While some options are being researched for their potential to support neurological function and mental health, the complexity of the brain and the current state of research necessitate a cautious and informed approach. It is essential for individuals with psychiatric or neurological conditions to consult closely with their healthcare providers to understand the potential impacts of these therapies on their condition and to monitor for any changes in symptoms or medication interactions.

Given the diversity of these treatments and their wide range of effects, individuals with a history of allergic reactions or sensitivities should proceed with caution. Allergic reactions to such therapies, though rare, can occur and may range from mild skin irritations to more severe systemic responses. Prior to initiating therapy, it is advisable for individuals with known allergies to undergo allergy testing when possible and to start with lower doses under medical supervision. Monitoring for any signs of allergic reaction during the initial stages of treatment is crucial for ensuring safety and for adjusting plans as necessary.

The decision to incorporate these therapies into health and wellness routines should be made with a comprehensive understanding of one's health history and current conditions. Collaborating with healthcare professionals who can provide personalized advice and monitoring is paramount. This collaborative approach ensures that therapy is not only tailored to the individual's health goals but also navigated in a manner that prioritizes safety and efficacy. By acknowledging and addressing these contraindications, individuals can better harness the potential benefits of these treatments while minimizing the risks associated with their use.

Chapter 11: Peptide Legality and Regulation

Peptide Regulations in the USA

In the United States, the regulation of peptides falls under the jurisdiction of the Food and Drug Administration (FDA) and the Drug Enforcement Administration (DEA), depending on the classification of the peptide as either a drug or a controlled substance. The FDA oversees the approval, marketing, and distribution of peptide drugs, ensuring they meet safety and efficacy standards for medical use. Peptides intended for therapeutic use must undergo rigorous clinical trials to demonstrate their safety and effectiveness before receiving FDA approval. This process involves several phases of research, starting with preclinical studies, followed by three phases of clinical trials involving human participants. Once a peptide drug has been proven safe and effective, it can be approved for marketing and prescribed by healthcare professionals.

Controlled substances, including certain peptides that mimic the effects of controlled drugs, are regulated by the DEA. These peptides may have restrictions on their manufacture, distribution, and possession to prevent misuse. For example, peptides that can potentially be used for performance enhancement in sports are closely monitored. The Anabolic Steroid Control Act of 1990 and its subsequent amendments have placed certain peptide hormones under the category of controlled substances, making their non-prescription use illegal.

The legal landscape for peptides also involves regulations on compounding pharmacies, which are often responsible for synthesizing peptide medications. These pharmacies must comply with state and federal regulations, including the Drug Quality and Security Act (DQSA), which outlines standards for compounding practices to ensure the quality and safety of compounded drugs. The FDA has issued guidance for compounding pharmacies that produce peptide medications, emphasizing the importance of sterile compounding techniques and accurate labeling.

Research peptides sold for laboratory use are another area of regulatory concern. These peptides are not intended for human use and are often labeled as such. However, there exists a gray market where research peptides are purchased for personal use, bypassing regulatory oversight intended to protect consumer safety. The FDA has issued warnings about the risks associated with the use of unapproved peptides, highlighting potential health hazards.

For consumers interested in peptide therapy, it's crucial to obtain these substances through legitimate medical channels. This means consulting with a healthcare provider who can prescribe FDA-approved medications or oversee treatment involving compounded formulations from reputable pharmacies. Consumers should be wary of these products sold online without a prescription, as they may not meet quality and safety standards.

Purchasing peptides for personal use requires due diligence to ensure compliance with legal and regulatory standards. When seeking peptide therapies, individuals should:

1. Consult with a healthcare professional to discuss the potential benefits and risks of peptide therapy.
2. Obtain peptides from reputable sources, such as licensed pharmacies that comply with FDA and DEA regulations.
3. Be cautious of peptides marketed as dietary supplements or research chemicals, as these may not be approved for human use and could pose health risks.
4. Stay informed about the legal status of specific peptides, especially those classified as controlled substances, to avoid legal repercussions.

Regulatory oversight aims to ensure that peptide therapies are safe, effective, and used responsibly. As research continues to uncover new therapeutic applications for peptides, regulatory frameworks may evolve to accommodate emerging treatments while safeguarding public health. Consumers and healthcare providers play a critical role in adhering to legal and regulatory standards, contributing to the responsible advancement of peptide therapy.

Buying Peptides Safely and Legally

Ensuring the safe and legal acquisition of peptides requires a multifaceted approach, beginning with a clear understanding of the regulatory environment and a commitment to adhering to it. Individuals seeking peptide therapies should prioritize engaging with healthcare providers who are not only knowledgeable about peptides but also committed to prescribing them within the bounds of the law and medical ethics. This professional guidance is invaluable, as it comes with an understanding of the latest research, potential side effects, and the legal nuances of peptide use. Healthcare providers can prescribe FDA-approved peptide medications, ensuring that the products have been subjected to rigorous testing for safety and efficacy.

When the conversation shifts to acquiring peptides, it's crucial to source these compounds from reputable pharmacies. Licensed pharmacies are subject to regulatory oversight, ensuring that they comply with stringent quality control measures. These establishments often provide compounded peptide medications, which are tailored to individual needs but still require adherence to the Drug Quality and Security Act (DQSA). This act sets forth standards for the practice of compounding, aiming to protect patients by ensuring that compounded medications are safe and effective. It's worth noting that while compounding pharmacies offer a degree of customization not available in mass-produced pharmaceuticals, this benefit comes with the responsibility of ensuring that the pharmacy is compliant with both state and federal regulations.

For those considering these compounds not available through traditional pharmaceutical channels, the realm of research substances presents an alternative. However, this area is fraught with potential legal and safety pitfalls. Research substances are intended for laboratory use, not for human consumption, and are often labeled accordingly. Despite this, a gray market exists where these substances are sold for personal use. It's critical to understand that products purchased in this manner may not have been subject to the same quality control measures as those intended for therapeutic use. This lack of oversight can lead to items that are impure, improperly labeled, or otherwise unsafe. The FDA has issued warnings about the risks associated with the use of unapproved substances, underscoring the importance of proceeding with caution.

To mitigate these risks, individuals interested in therapy should take proactive steps to verify the legitimacy and quality of the compounds they intend to purchase. This can involve requesting documentation of purity and composition, seeking out reviews and testimonials from other patients, and consulting with healthcare providers to interpret this information. It's also advisable to be cautious of substances marketed as dietary supplements or those that are ambiguously labeled as suitable for human consumption without the requisite approval.

Legal considerations extend beyond the point of purchase. The classification of certain compounds as controlled substances means that unauthorized possession or use can have legal repercussions. Staying informed about the legal status of specific substances is crucial to avoid inadvertently violating laws. This knowledge can also guide conversations with healthcare providers, ensuring that any prescribed items are not only effective for the intended use but also compliant with current regulations.

The landscape of therapy is evolving, with ongoing research continually uncovering new potential applications. As this field grows, so too does the complexity of navigating it safely and legally. The commitment to doing so, however, remains a constant. By prioritizing legal compliance and safety, individuals can explore the benefits of therapy while minimizing the risks, ensuring that their journey toward health and wellness is both effective and responsible.

Buying Peptides: Quality and Safety Tips

When purchasing peptides, the primary focus should be on quality and safety to ensure that the products will not only be effective but also not cause harm. It is crucial to verify the **purity** of the peptide, which refers to the percentage of the peptide in the product without any contamination. High purity levels, typically above **98%**, suggest that the product is of good quality and has been properly synthesized and processed. Another critical factor is the **source** of the peptide. Reputable suppliers will provide detailed information about the origin of their products, including manufacturing processes and quality control measures. It's essential to choose suppliers that adhere to **Good Manufacturing Practices (GMP)**, as this indicates compliance with regulatory standards designed to ensure product quality and safety.

The **formulation** of the peptide is another important aspect to consider. Peptides can be available in various forms, including lyophilized powder, liquid, or as part of a compounded formulation. The form can affect the stability and bioavailability of the peptide, so understanding the best form for your needs is vital. Additionally, the **storage** conditions of the peptide play a significant role in maintaining its integrity. Peptides may require refrigeration or protection from light to prevent degradation, so it's important to verify that the supplier has adhered to proper storage protocols before the product reaches you.

Documentation is a key element in verifying the quality and safety of peptides. Reputable suppliers should provide **Certificates of Analysis (CoA)** for their products, which include detailed information on purity, identity, and potency. The CoA should also provide information on any solvents, reagents, or other chemicals used during the synthesis process, as well as results from testing for contaminants such as bacteria, endotoxins, and heavy metals. It's advisable to request and review this documentation prior to making a purchase to ensure the peptides meet safety and quality standards.

Understanding the **peptide sequences** and **amino acid composition** is essential for ensuring that the product matches the intended therapeutic use. Incorrect sequences or amino acid substitutions can significantly alter the peptide's function and may lead to ineffective or harmful outcomes. Therefore, verifying the accuracy of the peptide sequence with the supplier is a critical step in the purchasing process.

Customer reviews and **testimonials** can provide insights into the experiences of others with the peptide and the supplier. While positive reviews can be encouraging, it's important to approach testimonials with a critical eye and consider the overall reputation of the supplier. Researching the supplier's history, including any regulatory actions or warnings from health authorities, can provide further assurance of their credibility and the quality of their products.

Finally, understanding the **legal and regulatory status** of peptides is crucial to ensure that your purchase complies with local laws and regulations. This includes knowing whether the peptide is approved for the intended use and if a prescription is required. Consulting with healthcare professionals or legal experts can provide guidance on these matters and help avoid potential legal issues.

In summary, when buying peptides, prioritizing quality and safety involves a comprehensive evaluation of the product, supplier, and legal considerations. By taking the time to thoroughly vet these aspects, individuals can make informed decisions that support their health and wellness goals while minimizing risks.

Safety and Reliable Peptide Sources

Identifying safe and reliable sources for these compounds becomes a paramount concern for those embarking on therapy, especially given the vast array of options and the varying degrees of quality and legality among suppliers. To ensure that one is engaging with a reputable source, several critical steps must be taken, each designed to vet the credibility and reliability of the supplier while ensuring the safety and efficacy of the substances acquired.

First and foremost, it's essential to seek out suppliers that have established a strong reputation within the healthcare and pharmaceutical communities. Suppliers that are widely recognized for their commitment to quality and safety are more likely to provide these compounds that meet stringent regulatory standards. Engaging in forums, reading reviews, and soliciting recommendations from healthcare professionals can provide insights into which suppliers are trusted within the community.

Moreover, transparency from the supplier about the sourcing, manufacturing, and testing of their products is crucial. Suppliers should be willing and able to provide detailed documentation on their offerings, including Certificates of Analysis (CoA) that verify the purity, potency, and safety of their substances. These documents should detail the results of rigorous testing for contaminants and impurities, ensuring that the products are of the highest quality. A lack of willingness to provide this documentation should be viewed as a red flag

Additionally, it's advisable to select suppliers that adhere to Good Manufacturing Practices (GMP), a set of guidelines enforced by regulatory agencies like the FDA. GMP certification indicates that the supplier's manufacturing processes meet the highest standards for quality and safety, significantly reducing the risk of contamination or inconsistency in the products offered.

Another important consideration is the level of expertise and knowledge demonstrated by the supplier. Suppliers that employ or consult with medical professionals and scientists are more likely to have a deep understanding of the compounds they offer, including their mechanisms of action, potential benefits, and risks. This expertise can be invaluable in guiding consumers toward the most appropriate substances for their needs, ensuring that they receive products that are not only safe but also effective for their intended use.

It's also critical to consider the legal and regulatory status of these compounds when selecting a supplier. Suppliers should be knowledgeable about the legal framework surrounding these substances, including which compounds are approved for therapeutic use and which are considered controlled substances. This knowledge is essential for ensuring that the products supplied are not only effective but also legal for use in the intended jurisdiction.

Engaging with suppliers that offer robust customer support is another key factor in ensuring a safe and positive experience with therapy. Suppliers that provide comprehensive support, including answering questions about product selection, usage, and potential side effects, demonstrate a commitment to customer safety and satisfaction. This support can be crucial in navigating the complexities of therapy, particularly for those new to this form of treatment.

In the realm of therapy, where the stakes involve not just legal compliance but also personal health and safety, the importance of selecting a reputable, reliable supplier cannot be overstated. By prioritizing suppliers that meet these stringent criteria, individuals can confidently embark on their therapy journey, assured in the knowledge that they are using high-quality, safe, and legal substances to support their health and wellness goals.

Chapter 12: Peptides in Nutrition and Lifestyle

Diet to Support Peptide Benefits

To maximize the benefits of peptide therapy, integrating certain dietary practices can significantly enhance peptide effectiveness. **Protein-rich foods** are fundamental in supporting the body's amino acid pool, which is necessary for the synthesis and function of peptides. Foods such as chicken, fish, beans, and nuts should be staples in your diet. These protein sources provide the body with a variety of amino acids, including those that are essential for peptide function and muscle recovery.

Antioxidant-rich fruits and vegetables also play a crucial role in supporting peptide therapy. Antioxidants help combat oxidative stress, which can damage cells and hinder the body's ability to repair itself. Incorporating a variety of colorful fruits and vegetables like berries, oranges, spinach, and carrots can provide a wide range of antioxidants, aiding in cellular repair and rejuvenation.

Healthy fats, particularly omega-3 fatty acids found in fish like salmon and plant sources like flaxseeds, are essential for reducing inflammation. This is particularly important for those using peptides for muscle recovery and anti-aging purposes. Omega-3s support cellular membrane health, which is crucial for the effective communication between cells and the optimal function of peptides.

Hydration cannot be overstated in its importance. Water plays a key role in maintaining cellular health and facilitating the transport of peptides throughout the body. Ensuring adequate hydration helps to optimize the cellular environment, allowing peptides to function more effectively.

Complex carbohydrates from whole grains, vegetables, and fruits provide the necessary energy for the body's metabolic processes. They also aid in the maintenance of stable blood sugar levels, which is important for minimizing insulin spikes that can interfere with peptide effectiveness. Foods like quinoa, sweet potatoes, and apples are excellent sources of complex carbohydrates that support overall health and wellness.

To further support peptide therapy, **limiting sugar and processed foods** is advisable. These foods can lead to inflammation and oxidative stress, which not only detracts from the benefits of peptides but can also contribute to aging and disease. Focusing on whole, unprocessed foods can enhance the body's response to peptide therapy.

Incorporating **fermented foods** such as yogurt, kefir, and kimchi into your diet can also benefit peptide therapy. These foods are rich in probiotics, which support gut health. A healthy gut microbiome is essential for the proper absorption and utilization of nutrients that support peptide function.

Lastly, **timing of nutrient intake** can also influence peptide effectiveness. Consuming protein and carbohydrates post-workout can aid in muscle recovery, while antioxidants consumed throughout the day can continuously support cellular health. Tailoring your nutrient intake to your daily activities and peptide administration schedule can further optimize the benefits of peptide therapy.

By focusing on a diet that supports cellular health, inflammation reduction, and overall wellness, individuals can enhance the effectiveness of peptide therapy. This holistic approach to health, combining targeted peptide use with supportive nutrition, can lead to significant improvements in muscle recovery, anti-aging efforts, and mental clarity.

Essential Nutrients for Peptide Benefits

The inclusion of micronutrients in one's diet cannot be overstated when discussing the optimization of peptide therapy for health and wellness. Micronutrients, comprising vitamins and minerals, play pivotal roles in cellular processes that are fundamental for the body's response to peptide treatment. For instance, Vitamin C, a potent antioxidant, not only aids in the neutralization of free radicals but also is essential in the synthesis of collagen, a peptide that ensures skin elasticity and wound healing. The consumption of citrus fruits, strawberries, bell peppers, and dark leafy greens can provide the body with ample Vitamin C, thus supporting skin health and anti-aging efforts facilitated by peptide therapy.

Similarly, Vitamin D, often referred to as the "sunshine vitamin," enhances the body's absorption of calcium and phosphorus, minerals critical for bone health. This is particularly relevant for individuals utilizing peptides for muscle recovery and performance, as strong bones form the foundation upon which muscles operate. Fatty fish like salmon and mackerel, egg yolks, and fortified foods are rich sources of Vitamin D, and exposure to sunlight also naturally boosts Vitamin D levels in the body.

Zinc, a trace mineral, is another essential nutrient that supports the immune system and plays a role in cellular division and repair. Its importance in peptide therapy lies in its ability to facilitate the synthesis of protein and thus peptides within the body. Foods high in zinc such as oysters, beef, pumpkin seeds, and lentils, when incorporated into the diet, can enhance the effectiveness of peptide treatments aimed at tissue repair and regeneration.

Magnesium, involved in over 300 enzymatic reactions in the body, including protein synthesis, is crucial for those undergoing peptide therapy for muscle recovery and mental clarity. It aids in muscle relaxation and reduces cramping, which can be beneficial post-exercise. Nuts, seeds, whole grains, and leafy green vegetables are excellent sources of magnesium.

The role of essential fatty acids, particularly omega-3s, extends beyond their anti-inflammatory properties. They are integral to the structure of cell membranes, affecting the function of cell receptors. These fatty acids can facilitate the binding of peptides to their respective receptors, thereby enhancing the signaling pathways that lead to improved muscle recovery, brain health, and anti-aging effects. Flaxseeds, chia seeds, walnuts, and algae oil are plant-based sources of omega-3 fatty acids, offering an alternative for those who do not consume fish.

Hydration, while not a nutrient per se, remains a cornerstone of effective peptide therapy. Water is essential for the optimal functioning of every cell in the body and plays a critical role in the transport and distribution of nutrients, including peptides. Adequate hydration ensures that the tissues are well-saturated, facilitating the efficient delivery of peptides to their target sites within the body.

In the context of enhancing peptide benefits through nutrition, it is imperative to consider not just what to eat but also when and how much. Nutrient timing, particularly around exercise, can influence the body's physiological response to peptide therapy. For instance, consuming protein and carbohydrates post-exercise can enhance muscle repair and growth, a process further supported by peptide therapy. Similarly, the strategic intake of antioxidants throughout the day can provide ongoing support against oxidative stress, complementing the anti-aging effects of peptide treatments.

By focusing on a diet rich in essential nutrients, individuals can create a conducive environment that maximizes the therapeutic effects of peptides. This holistic approach, integrating targeted nutrition with peptide therapy, holds the potential to significantly amplify outcomes related to muscle recovery, anti-aging, and overall wellness.

Exercise and Rest Regime

Exercise and rest form the cornerstone of any comprehensive health and wellness strategy, especially when integrating peptides into one's routine. The symbiotic relationship between physical activity and recovery periods is paramount for maximizing the benefits of peptide therapy. Engaging in regular exercise stimulates the body's natural production of growth hormone, which peptides can further enhance, leading to improved muscle recovery, fat loss, and overall well-being. However, it is crucial to balance this physical activity with adequate rest and recovery to prevent overtraining and allow the body to repair and strengthen.

Exercise Recommendations: To leverage the full potential of peptides, incorporating a mix of cardiovascular, strength, and flexibility training into your weekly routine is advisable. Cardiovascular exercises, such as brisk walking, running, or cycling, enhance circulatory health and aid in the efficient distribution of peptides throughout the body. Strength training, on the other hand, is essential for building and maintaining muscle mass, a process that peptides can significantly support by promoting tissue repair and growth. Flexibility exercises, including yoga or Pilates, contribute to overall mobility and help in preventing injuries, ensuring that one can continue to exercise regularly without prolonged interruptions.

Rest and Recovery: Equally important to the exercise regimen is rest. The body repairs and strengthens itself in the time between workouts, not during the exercises themselves. Therefore, ensuring sufficient sleep, typically 7-9 hours per night for most adults, is critical. During sleep, growth hormone levels naturally increase, which peptides can further enhance, thereby supporting the body's healing processes. Incorporating rest days into your exercise schedule allows muscles to recover and rebuild, reducing the risk of injury. Active recovery, such as light walking or gentle stretching, can also be beneficial by promoting circulation and flexibility without overexerting the body.

Periodization and Peptide Therapy: Tailoring your exercise and rest regime to align with your peptide therapy can optimize results. Periodization, or structuring your training and recovery phases, can be synchronized with peptide cycles to enhance muscle growth, fat loss, or recovery, depending on your goals. For instance, during periods of intensive training, increasing peptide dosage, as advised by a healthcare professional, can support muscle recovery and growth. Conversely, during recovery or tapering phases, adjusting peptide types or dosages can aid in healing and prepare the body for the next phase of training.

Listening to Your Body: It's imperative to listen to your body's signals. Overtraining can lead to fatigue, decreased performance, and increased risk of injury, which can be counterproductive to your health goals. If you experience prolonged soreness, fatigue, or other signs of overtraining, it may be necessary to adjust your exercise intensity, duration, or frequency. Similarly, peptides are not a one-size-fits-all solution; their effectiveness can vary based on individual factors such as age, health status, and lifestyle. Regular consultation with healthcare professionals can ensure that your peptide regimen and exercise routine are appropriately matched to your body's needs, promoting optimal health and performance.

Hydration and Nutrition: Supporting your exercise and rest regime with proper hydration and nutrition is essential. Water assists in nutrient transport and helps maintain optimal cellular function, which is crucial for the effectiveness of peptide therapy. A balanced diet rich in proteins, healthy fats, and carbohydrates fuels the body for exercise and aids in recovery, while vitamins and minerals support overall health and the body's response to peptide therapy.

In conclusion, integrating an informed and balanced approach to exercise and rest with peptide therapy can significantly enhance health outcomes. By carefully planning your physical activity, ensuring adequate recovery, and tailoring peptide usage to support these efforts, individuals can achieve improved muscle recovery, anti-aging effects, and overall wellness.

Fitness, Sleep, and Peptide Effectiveness

Understanding the intricate balance between fitness, sleep, and the effectiveness of these bioactive compounds is crucial for those integrating them into their health and wellness routines. The physiological impacts of exercise and rest on their activity are profound, influencing how they are synthesized, function, and ultimately, how they contribute to health outcomes. The interplay between these elements can enhance the overall effectiveness of therapy, particularly when considering the goals of anti-aging, muscle recovery, and mental clarity.

Physical activity plays a pivotal role in the body's hormonal balance, significantly affecting growth hormone levels, which are essential for the synthesis and function of these compounds. Regular, structured exercise routines can increase the body's natural production of these hormones, creating an optimal environment for them to exert their beneficial effects. This activity-induced boost in growth hormone levels not only facilitates muscle recovery and growth but also enhances the body's ability to utilize these substances for tissue repair and regeneration. However, the type, intensity, and duration of exercise are critical factors to consider. Overexertion without adequate rest can lead to elevated cortisol levels, potentially counteracting the positive effects of therapy by inducing a state of physiological stress rather than recovery.

Adequate sleep is another cornerstone of maximizing effectiveness. Sleep is a critical period for the body's repair processes, with growth hormone levels naturally peaking during deep sleep cycles. This increase in growth hormone during sleep complements therapy by enhancing cellular repair and regeneration processes. The quality and quantity of sleep directly affect the body's ability to recover, with poor sleep patterns potentially diminishing the effectiveness of these compounds. Ensuring a consistent sleep schedule of 7-9 hours per night can significantly amplify the rejuvenating effects of therapy, supporting anti-aging efforts, muscle recovery, and cognitive function.

The synergy between exercise, sleep, and this therapy is a dynamic that requires careful management. Tailoring exercise routines to include a mix of cardiovascular, strength, and flexibility training, while ensuring adequate rest and recovery periods, can optimize the body's response to treatment. This includes considering the timing of administration in relation to exercise and sleep schedules. For example, administering certain substances before sleep may leverage the natural spike in growth hormone during the night, enhancing the therapeutic effects. Similarly, the use of compounds that support muscle recovery can be timed around exercise sessions to maximize tissue repair and growth.

Furthermore, the impact of hydration and nutrition cannot be overlooked in this context. Proper hydration is essential for maintaining optimal blood flow and nutrient delivery, crucial for the effective distribution and function of these bioactive substances within the body. A diet rich in proteins, healthy fats, and antioxidants supports the repair, recovery, and rejuvenation processes facilitated by therapy. Nutritional timing also plays a role, with certain nutrients potentially enhancing the effectiveness of these compounds when consumed in proximity to exercise or sleep.

In essence, the integration of these bioactive substances into a health and wellness routine is not a standalone solution but rather a component of a holistic approach that includes balanced exercise, adequate rest, and targeted nutrition. Understanding and applying the principles of how fitness and sleep impact effectiveness can lead to significant improvements in health outcomes, particularly in the realms of anti-aging, muscle recovery, and mental clarity. Each element—exercise, sleep, nutrition, and therapy—works in concert to support the body's natural healing and rejuvenation processes, underscoring the importance of a comprehensive, informed approach to health and wellness.

Complete Wellness Strategies

Mental wellness plays a crucial role in the holistic approach to health, especially when integrating peptides into one's wellness routine. The mind-body connection is undeniable, with stress, anxiety, and poor mental health having profound effects on physical well-being and vice versa. Therefore, incorporating practices that promote mental clarity and reduce stress is essential for maximizing the benefits of peptide therapy. Mindfulness meditation, for example, has been shown to reduce stress, improve concentration, and enhance overall well-being. By dedicating a few minutes each day to mindfulness or meditation practices, individuals can create a more conducive environment for the body to respond to peptide therapy, promoting healing and rejuvenation from within.

Another important aspect of mental wellness is ensuring adequate social support and engagement. Social interactions and relationships can have a significant impact on health, influencing stress levels, mood, and even the body's physical response to treatments. Engaging in community activities, maintaining close relationships, and seeking support when needed can enhance the effectiveness of peptide therapy by fostering a positive mental state.

Sleep hygiene is another critical component of a complete wellness strategy. The quality of sleep directly impacts the body's repair processes, hormone regulation, and mental health. Practices that promote good sleep hygiene, such as maintaining a consistent sleep schedule, creating a restful environment, and avoiding stimulants before bedtime, can significantly improve the effectiveness of peptide therapy. Good sleep hygiene not only supports the body's natural healing processes but also enhances cognitive function and mood, contributing to a more holistic approach to health and wellness.

Environmental factors also play a role in holistic health, with exposure to natural environments having been shown to reduce stress, improve mood, and enhance physical well-being. Spending time outdoors, whether through exercise, gardening, or simply taking walks in nature, can complement peptide therapy by promoting relaxation and reducing stress. The natural environment provides a unique set of stimuli that can help reset the body's stress response, improving overall health and well-being.

Finally, the role of lifelong learning and cognitive engagement in maintaining brain health cannot be overlooked. Engaging in activities that challenge the mind, such as learning a new language, playing musical instruments, or participating in puzzles and brain games, can help maintain cognitive function and mental clarity. This cognitive engagement is particularly important as we age, with research suggesting that lifelong learning can help delay cognitive decline. Integrating cognitive activities into one's routine, alongside peptide therapy, can provide a comprehensive approach to health that supports both the body and mind.

By addressing mental wellness, sleep hygiene, environmental exposure, and cognitive engagement, individuals can adopt a truly holistic approach to health that maximizes the benefits of peptide therapy. These strategies, when combined with a balanced diet, regular exercise, and targeted peptide use, can lead to significant improvements in overall well-being, supporting both physical and mental health in a comprehensive manner.

Integrating Peptides into Health Routines

Integrating these compounds into a holistic health routine requires a nuanced understanding of their role within the broader context of wellness. This approach emphasizes the synergy between these substances and lifestyle factors such as diet, exercise, sleep, and mental health practices. To optimize the benefits of these compounds, one must consider them not as isolated agents but as components of a comprehensive wellness strategy. This strategy should be personalized, taking into account individual health goals, existing conditions, and lifestyle preferences.

The first step in incorporating these compounds into a health routine is to establish clear, evidence-based objectives. Whether the aim is to enhance muscle recovery, support anti-aging efforts, or improve mental clarity, each goal will dictate a specific protocol as well as complementary lifestyle adjustments. For instance, substances known for their muscle recovery properties may be most beneficial when coupled with a protein-rich diet and a structured exercise regimen that includes both resistance and flexibility training. Similarly, compounds with anti-aging benefits can be paired with a skincare routine rich in antioxidants and hydration, alongside dietary choices that support skin health.

Consultation with healthcare professionals is essential to tailor therapy to individual needs. This personalized approach ensures not only the selection of the most appropriate substances but also the determination of optimal dosages and administration methods. Subcutaneous injections, for example, may offer higher bioavailability for certain compounds, while oral supplements or nasal sprays might be preferred for their convenience and ease of use. The method of administration should align with the individual's lifestyle, ensuring that the regimen is both effective and sustainable.

Monitoring and adjustment are key components of integrating these compounds into a holistic health routine. Regular check-ins with healthcare providers allow for the assessment of progress and the identification of any necessary adjustments to the protocol. This iterative process ensures that the regimen remains aligned with the individual's evolving health goals and responds to any changes in their condition or lifestyle. Keeping a detailed journal of use, including dosages, administration times, and any side effects or benefits observed, can provide valuable insights that inform these adjustments.

The role of education in the successful integration of these compounds into a health routine cannot be overstated. A solid understanding of how these substances function, their potential benefits and risks, and the importance of factors such as timing and synergy with other wellness practices empowers individuals to make informed decisions about their health. This knowledge base should be built through reputable sources, including scientific literature, consultations with healthcare professionals, and educational resources provided by trusted organizations in the field of therapy.

Beyond the biochemical impact of these compounds, their use should be viewed within the larger framework of a health-conscious lifestyle. This includes not only diet and exercise but also stress management, sleep quality, and environmental factors that influence well-being Practices such as mindfulness meditation, yoga, and spending time in nature can enhance the body's response to therapy by reducing stress and improving overall health. The integration of these substances into such a routine underscores the interconnectedness of physical, mental, and emotional well-being, highlighting the holistic nature of true health optimization.

In aligning the use of these compounds with a holistic health routine, it is crucial to consider not just the immediate effects of therapy but also the long-term implications for health and wellness. The goal is to support the body's natural processes and promote resilience and vitality through a balanced approach to wellness. This requires a commitment to ongoing learning, adaptation, and self-care, with these substances serving as one of many tools in the pursuit of optimal health. By carefully considering how these compounds fit into this broader context, individuals can maximize their benefits while minimizing risks, leading to sustainable improvements in health and quality of life.

Chapter 13: Testimonials and Case Studies

Real Experiences with Peptides

In the realm of health and wellness, peptides have emerged as a beacon of hope for many, offering solutions to problems that seemed insurmountable. The stories of those who have experienced the transformative power of peptides are not just testimonials; they are real-life affirmations of the science-backed benefits discussed throughout this book. One such story comes from Emily, a 42-year-old marathon runner who struggled with chronic inflammation and joint pain, threatening to end her passion for running. After incorporating **BPC-157**, a peptide known for its remarkable healing properties, into her regimen, Emily noticed a significant reduction in pain and inflammation, allowing her to continue running without discomfort. Her recovery times improved, and she was able to increase her training intensity, ultimately achieving personal best times in her races.

Another compelling case is that of Sarah, a 35-year-old executive who faced the relentless demands of her high-stress job. The toll on her mental clarity and energy levels was palpable until she discovered the cognitive-enhancing effects of **Semax**. This peptide, known for its neuroprotective properties, helped Sarah regain her sharpness and focus, enabling her to tackle her workload with renewed vigor and efficiency. Moreover, the improvement in her stress levels was an unexpected yet welcome benefit, illustrating the multifaceted advantages of peptides.

The journey of Mark, a 50-year-old who noticed the inevitable signs of aging on his skin, highlights the cosmetic benefits of peptides. After incorporating a skincare regimen that included **Matrixyl** and **Copper peptides**, Mark observed a noticeable improvement in his skin's texture and elasticity. The depth of his wrinkles diminished, and his skin appeared more radiant and youthful. This transformation was not just skin deep; it boosted his confidence and satisfaction with his appearance.

These stories underscore the practical and profound impact peptides can have on various aspects of health and wellness. From enhancing physical performance and recovery to improving cognitive function and skin health, the potential applications are vast and varied. The experiences of Emily, Sarah, and Mark are a testament to the power of peptides, offering a glimpse into the possibilities that lie ahead for those willing to explore this promising frontier in health science. Their successes are not anomalies but rather examples of the achievable benefits when peptides are used thoughtfully and responsibly, guided by the principles outlined in this book.

Success Stories in Health Improvement

Building on the momentum of the transformative experiences shared by Emily, Sarah, and Mark, the narrative of these compounds' impact on health and wellness continues to unfold through the story of Alex, a 45-year-old avid cyclist who faced a plateau in his performance and recovery times. The introduction of these substances specifically tailored for endurance and recovery marked a turning point in his training regimen. Utilizing a combination of IGF-1 for its muscle-building properties and MGF for its role in muscle repair, Alex experienced a notable improvement in his stamina and a decrease in recovery periods between intense cycling sessions. This adjustment allowed him to surpass his previous performance thresholds, demonstrating the efficacy of these compounds in enhancing physical endurance and recovery in sports enthusiasts.

Further illustrating the versatility of these substances, the case of Jenna, a 38-year-old mother of two and a part-time yoga instructor, sheds light on the holistic benefits beyond the realms of physical performance and cosmetic enhancement. Struggling with postpartum depression and fatigue, Jenna found solace in Semax and Selank, known for their mood-enhancing and anxiolytic effects. The incorporation of these substances into her wellness routine led to a significant improvement in her mental health, providing a sense of balance and well-being that had been elusive. Jenna's experience underscores the potential of these compounds to support mental and emotional health, offering a beacon of hope for those grappling with similar challenges.

The narrative of these substances as a catalyst for health improvement extends to the realm of metabolic health through the experience of Tom, a 47-year-old with a predisposition to type 2 diabetes. Facing the reality of his condition, Tom sought out compounds known for their ability to regulate blood sugar levels and enhance insulin sensitivity. The introduction of Ipamorelin and CJC-1295, with growth hormone-releasing properties, into his regimen not only helped stabilize his blood sugar levels but also facilitated a reduction in visceral fat, a common challenge among individuals with diabetes. Tom's journey highlights the role of these substances in managing and potentially mitigating the effects of metabolic disorders, showcasing their utility in addressing a broad spectrum of health concerns.

These narratives collectively paint a picture of the profound and multifaceted impact that these compounds can have on health and wellness. From enhancing physical performance and recovery to supporting mental health and managing metabolic disorders, the stories of Emily, Sarah, Mark, Alex, Jenna, and Tom offer compelling evidence of the transformative potential of these substances. Each story, unique in its focus and outcomes, contributes to the broader understanding of how these compounds can be integrated into health and wellness strategies to address specific concerns and improve overall quality of life. The experiences detailed here not only validate the scientific principles discussed throughout the book but also provide real-world testimonials to the practical benefits, reinforcing their value as a pivotal component of contemporary health and wellness regimens.

Clinical Case Studies on Peptide Efficacy

Delving deeper into the scientific underpinnings of peptide therapy, it's essential to highlight the case of Jonathan, a 48-year-old firefighter, who suffered from severe burns during a rescue operation. The integration of **GHK-Cu**, a peptide renowned for its wound healing and anti-inflammatory properties, played a pivotal role in Jonathan's recovery. The application of GHK-Cu significantly accelerated wound healing, reduced scarring, and improved the overall skin texture in the affected areas. This case not only demonstrates the regenerative capacity of peptides in skin repair but also underscores the broader implications for trauma recovery and regenerative medicine.

In another instance, the focus shifts to the metabolic efficiency facilitated by peptide therapy. Linda, a 54-year-old diabetic, struggled with fluctuating blood glucose levels and insulin resistance, a common challenge in the management of type 2 diabetes. The introduction of **GLP-1** (Glucagon-like Peptide-1) analogs into her treatment regimen marked a significant turning point. These peptides, which mimic the action of the endogenous hormone GLP-1, helped improve Linda's insulin sensitivity, promote satiety, and facilitate weight loss. Her case is a testament to the potential of peptides in modulating metabolic pathways, offering a promising adjunctive therapy for individuals with metabolic disorders.

The therapeutic potential of peptides extends into the realm of autoimmune diseases, as illustrated by the case of Michael, a 37-year-old diagnosed with multiple sclerosis (MS). The administration of **Thymosin Beta-4** (TB-500), known for its immunomodulatory and tissue repair properties, provided remarkable benefits. Michael experienced a reduction in the frequency of MS flare-ups, alongside improvements in muscle strength and neurological function. This case highlights the immunoregulatory and neuroprotective effects of peptides, suggesting a novel approach to managing autoimmune conditions and enhancing neuroregeneration.

Furthermore, the impact of peptides on cardiovascular health is exemplified by the case of Angela, a 62-year-old with a history of hypertension and atherosclerosis. The use of **Angiotensin (1-7)**, a peptide with vasodilatory and anti-inflammatory effects, contributed to a notable improvement in her cardiovascular profile. Angela reported lower blood pressure levels and a reduction in arterial stiffness, indicative of improved vascular health. This case underscores the potential of peptides in cardiovascular disease management, emphasizing their role in vascular protection and endothelial function.

The diverse applications of peptides in health and wellness are further illustrated by the case of Derek, a 29-year-old athlete recovering from an anterior cruciate ligament (ACL) injury. Incorporating **BPC-157** and **MGF** (Mechano Growth Factor) into his rehabilitation protocol expedited his recovery by enhancing ligament healing and muscle regeneration. Derek's swift return to competitive sports highlights the role of peptides in sports medicine, offering evidence of their efficacy in tissue repair and recovery following injury.

These clinical case studies collectively illuminate the broad spectrum of peptide therapy applications, from enhancing metabolic health and managing autoimmune diseases to promoting tissue regeneration and improving cardiovascular health. Each case provides a unique insight into the practical benefits of peptides, supported by scientific evidence and real-life outcomes. The experiences of Jonathan, Linda, Michael, Angela, and Derek not only validate the therapeutic potential of peptides but also reinforce the importance of a personalized approach to peptide therapy, tailored to meet the specific needs and health objectives of each individual.

Peptide Health Improvement Cases

The utilization of peptides in addressing complex health issues extends into the management of chronic pain, a condition that plagues millions of individuals worldwide. Chronic pain, often debilitating and resistant to conventional treatments, presents a significant challenge in the medical field. The case of Elizabeth, a 58-year-old woman suffering from fibromyalgia, a condition characterized by widespread musculoskeletal pain accompanied by fatigue, sleep, memory, and mood issues, illuminates the role of peptides in pain management. Elizabeth found relief through the administration of Palmitoylethanolamide (PEA), a fatty acid amide belonging to the endocannabinoid family, which has been shown to exert a natural anti-inflammatory and pain-relieving effect. PEA, through its action on the endocannabinoid system, helped to significantly reduce Elizabeth's pain perception, improving her quality of life and allowing her to engage in daily activities with renewed vigor. This case not only highlights the potential of peptides in managing pain but also points to their ability to target underlying mechanisms of chronic conditions, offering a beacon of hope for those who have found little relief through traditional avenues.

The spectrum of peptide therapy also encompasses the treatment of gastrointestinal disorders, illustrating the versatility and breadth of peptide applications in health and wellness. Consider the case of James, a 47-year-old male diagnosed with inflammatory bowel disease (IBD), a term that covers a number of chronic inflammatory gastrointestinal disorders, including Crohn's disease and ulcerative colitis. James experienced significant improvement in his condition after the introduction of BPC-157, a peptide known for its potent anti-inflammatory and healing properties. BPC-157, administered through subcutaneous injections, played a critical role in reducing gastrointestinal inflammation, promoting healing of intestinal lesions, and restoring gut health. James' case underscores the therapeutic potential of peptides in treating gastrointestinal conditions, offering a novel approach to managing diseases that are often resistant to conventional treatments.

Furthermore, the role of these bioactive compounds in enhancing fertility presents another dimension of their therapeutic potential. The case of Sophia, a 34-year-old woman struggling with infertility, showcases the application of these substances in reproductive health. Sophia was introduced to a regimen that included the use of Gonadotropin-releasing hormone (GnRH) compounds, which are known to modulate the release of follicle-stimulating hormone (FSH) and luteinizing hormone (LH), both crucial for ovulation and fertility. The GnRH compounds helped to regulate Sophia's hormonal imbalances, ultimately leading to successful conception. This example highlights the intricate ways in which these substances can influence the endocrine system, offering solutions to complex reproductive health issues and opening doors to individuals and couples facing fertility challenges.

The therapeutic applications of these compounds further extend to the realm of mental health, where they offer innovative approaches to managing mood disorders and enhancing cognitive function. The case of Daniel, a 41-year-old software developer experiencing severe anxiety and depression, exemplifies the impact of these substances on mental health. Daniel found significant relief through the use of Selank, a compound with anxiolytic properties that modulate neurotransmitter levels and enhance the stress response. Selank provided a noticeable improvement in Daniel's symptoms, reducing anxiety and elevating mood without the side effects commonly associated with traditional anxiolytics and antidepressants. This case highlights the potential of these bioactive compounds as modulators of mental health, providing a promising alternative for those seeking to manage mood disorders and improve cognitive function.

In conclusion, the diverse and impactful examples of therapy in clinical case studies underscore the broad spectrum of their applications in health and wellness. From managing chronic pain, gastrointestinal disorders, and infertility to enhancing mental health, these substances offer a promising and innovative approach to addressing a wide range of health issues. These cases not only validate the therapeutic potential of these compounds but also highlight the importance of personalized medicine, where treatments are tailored to the unique needs and conditions of each individual. As research continues to unravel the mechanisms and benefits of these bioactive substances, their role in advancing healthcare and improving quality of life becomes increasingly evident, marking a new frontier in medical science and therapeutic intervention.

Conclusion

Final Thoughts on Peptide Use

The exploration of peptides in the realm of health and wellness is not just a testament to the advancements in medical science but also a beacon of hope for individuals striving for a holistic approach to health. The efficacy and importance of these biomolecules extend beyond the traditional boundaries of medicine, offering innovative solutions to age-old problems of aging, muscle recovery, and cognitive decline. As we delve deeper into the scientific intricacies of this therapy, it becomes evident that the potential for personalized health strategies is immense. These compounds, with their unique ability to mimic natural biological processes, represent a key to unlocking a new era of targeted treatments that cater to the specific needs of individuals. The diversity of these substances, from those enhancing muscle growth and repair to those improving skin health and cognitive function, underscores their versatility in addressing a wide range of health concerns.

The significance of these agents in health and wellness cannot be overstated. Their role in promoting cellular repair, regeneration, and overall homeostasis is a cornerstone of their therapeutic potential. For health-conscious individuals, particularly women aged between 30 and 50, the appeal lies in their ability to provide evidence-based solutions to the challenges of aging and lifestyle-related health issues. The stories of individuals who have experienced the transformative effects of this therapy illuminate the path for others seeking to enhance their quality of life through science-backed health interventions. These narratives not only provide a glimpse into the practical benefits but also offer a source of inspiration for those on the fence about incorporating these solutions into their wellness regimen.

The responsible use of these compounds, guided by scientific research and medical advice, is paramount. As with any therapeutic intervention, the principles of safety, efficacy, and ethical use apply. The landscape of this therapy is continually evolving, with ongoing research shedding light on new applications and optimizing existing protocols.

This dynamic field of study promises to bring forth novel agents and usage strategies, further expanding the toolkit available for health and wellness optimization. It is crucial for individuals to stay informed, consult with healthcare professionals, and consider the holistic implications of this therapy within the broader context of their health and lifestyle goals.

The integration of these substances into health and wellness strategies signifies a shift towards more personalized and precision-based approaches to healthcare. This paradigm shift, fueled by the convergence of biotechnology and medical science, opens up new avenues for enhancing human health and longevity. As we stand on the cusp of this exciting frontier, the possibilities for improving quality of life through this therapy are vast and varied. The journey of exploring and utilizing these agents is an ongoing process of discovery, learning, and adaptation, driven by the collective pursuit of optimal health and wellness. The role of these compounds in this journey is undeniably significant, offering a powerful means to address specific health concerns while contributing to the overarching goal of achieving lasting wellness.

Holistic Long-Term Approach

Adopting a holistic, long-term approach to health and wellness through the use of peptides requires an understanding of how these powerful molecules interact with various bodily systems. It's essential to recognize that peptides are not a standalone solution but rather a component of a broader strategy aimed at achieving and maintaining optimal health. This strategy encompasses a balanced diet, regular physical activity, adequate rest, and mental well-being, each playing a critical role in enhancing the efficacy of therapeutic compounds. Nutrition is foundational to any health regimen, with certain nutrients acting synergistically with these compounds to support cellular function and regeneration. A diet rich in proteins provides the amino acids necessary for synthesis and function, while antioxidants from fruits and vegetables can protect cells from oxidative stress, potentially enhancing the longevity and efficacy of treatments. Physical activity is another cornerstone of a holistic health approach, with exercise not only improving muscle mass and cardiovascular health but also potentially boosting the body's response to these compounds. Regular exercise can increase the sensitivity of receptors that these compounds bind to, making the body more receptive to their effects. Rest and recovery are equally important, as sleep is a critical period for the body's repair processes, during which these compounds can exert their regenerative effects most efficiently.

Furthermore, mental well-being plays a significant role in the holistic approach to health. Stress can have profound negative effects on the body, including reducing the effectiveness of therapeutic compounds in promoting healing and regeneration. Techniques such as mindfulness, meditation, and stress management practices can enhance the body's response to these therapies by creating a more conducive environment for healing and rejuvenation.

Incorporating these compounds into a lifestyle that values balance across these areas can amplify the benefits of therapy. For example, compounds that promote muscle recovery and growth will be more effective when combined with a protein-rich diet and a consistent exercise regimen. Similarly, compounds used for skin health will yield better results when the individual practices good skincare, stays hydrated, and protects their skin from excessive sun exposure.

It's also crucial to understand that the effectiveness of these compounds can vary widely among individuals, due to factors such as age, health status, lifestyle, and genetic predispositions. This variability underscores the importance of personalized protocols, developed in consultation with healthcare professionals, to ensure the most effective and safe use of these therapies. Regular monitoring and adjustment of regimens are necessary to adapt to the body's changing needs and responses over time.

Finally, the holistic, long-term approach to using these compounds for health and wellness emphasizes the principle of prevention over treatment. By supporting the body's natural processes and addressing potential health issues before they become problematic, these therapies can contribute to a proactive strategy for maintaining health and preventing disease. This approach aligns with the broader goals of enhancing quality of life and promoting longevity, making these compounds a valuable tool in the pursuit of holistic health and wellness.

Conscious Health Journey

As you stand at the threshold of incorporating peptides into your health and wellness routine, it is imperative to approach this new chapter with both enthusiasm and caution. The journey you are about to embark on is not just about the peptides themselves but about fostering a deeper understanding of your body's needs and how to meet them responsibly. Remember, the essence of this journey is not in seeking quick fixes but in nurturing lasting wellness through informed choices. The information provided throughout this book aims to equip you with the knowledge to make those choices with confidence.

It is crucial to start with a clear objective in mind, be it enhancing muscle recovery, supporting your mental clarity, or embarking on an anti-aging regimen. Each goal requires a tailored approach, considering the specific peptides that align with your objectives, the optimal dosages, and the most effective administration routes. However, beyond the specifics of peptide use, this journey is also about integrating these powerful tools into a balanced lifestyle that includes proper nutrition, regular exercise, adequate rest, and mindfulness practices.

Engaging with a healthcare professional who understands your health history and goals can provide invaluable guidance and support as you navigate the complexities of peptide therapy. This collaboration ensures not only the safety and effectiveness of your regimen but also its customization to your unique physiological makeup.

Moreover, as you proceed, maintaining a journal to track your progress, experiences, and any side effects can be incredibly insightful. This practice not only helps in fine-tuning your regimen but also in understanding your body's responses to different peptides and dosages.

Ultimately, the decision to integrate peptides into your health routine is a significant step towards self-empowerment. It signifies a commitment to not just living longer but living better. With the right approach, informed choices, and mindful integration, peptides can indeed be a transformative element in your quest for lasting wellness. Remember, this is not just a journey of health optimization but also one of self-discovery and growth.

Appendices

Glossary of Technical Terms

- **Bioavailability**: Refers to the proportion of a substance that enters the circulation when introduced into the body and so is able to have an active effect. This term is crucial when discussing the effectiveness of different peptide administration methods, such as injections, sprays, and supplements.
- **Collagen**: A protein that is a major component of connective tissues in the body. Peptides that stimulate collagen production are often used for improving skin health and reducing the signs of aging.
- **Cycles**: In the context of peptide therapy, this refers to the planned sequence of dosage and rest periods to maximize the therapy's effectiveness while minimizing potential side effects.
- **Dosage**: The amount of a substance to be taken at one time or the total daily intake, tailored according to individual factors such as body weight, goals, and tolerance.
- **GHRP (Growth Hormone Releasing Peptides)**: A class of peptides that stimulate the pituitary gland to release growth hormone, used for various purposes including muscle growth, fat loss, and anti-aging.
- **IGF-1 (Insulin-like Growth Factor-1)**: A protein similar in structure to insulin that plays an important role in childhood growth and continues to have anabolic effects in adults.
- **Lipolytic Peptides**: Peptides that help break down fats, often used in weight management protocols to accelerate fat loss.
- **Matrixyl (palmitoyl pentapeptide-4)**: A peptide used in skincare products to reduce the appearance of wrinkles and improve skin elasticity by promoting collagen production.
- **Peptide**: Short chains of amino acids, the building blocks of proteins, which can influence various bodily functions by mimicking or stimulating natural processes.

- **Senescence**: The process by which cells cease to divide and grow, contributing to aging and the development of age-related diseases. Certain peptides are studied for their potential to delay senescence and promote rejuvenation.

- **Subcutaneous Injection**: A method of administering peptides using a small needle to inject the substance into the fat layer beneath the skin, often preferred for its bioavailability and ease of self-administration.

Additional Resources

For those seeking to deepen their understanding of peptides and their application in health and wellness, a wealth of resources is available to complement the information presented in this book. **Articles** published in reputable scientific journals such as *The Journal of Peptide Science* and *Peptides* offer cutting-edge research findings and reviews on peptide synthesis, biology, and therapeutic applications. These peer-reviewed articles provide a solid scientific foundation for those interested in the biochemical mechanisms and potential health benefits of peptides.

Books that come highly recommended include *Peptides: The Wave of the Future*, which offers a comprehensive overview of peptide science from research to clinical applications. Another valuable resource is *The Peptide Handbook*, providing practical guidance on peptide use, including dosages, administration methods, and safety considerations. These texts serve as excellent references for both beginners and those with more advanced knowledge, offering detailed insights into peptide therapy's complexities.

Websites such as the American Peptide Society (www.americanpeptidesociety.org) offer a platform for education and networking among scientists, clinicians, and individuals interested in peptides. The website features a variety of resources, including conference proceedings, educational materials, and links to relevant research articles. Another useful online resource is www.peptidesciences.com, which not only supplies peptides but also provides a wealth of information on peptide research, including white papers and case studies that illustrate the practical applications of peptides in health and wellness.

By exploring these **articles, books, and websites**, readers can expand their knowledge and stay informed about the latest developments in peptide science. This additional reading will empower individuals to make educated decisions about incorporating peptides into their health regimen, ensuring they do so with a solid understanding of the benefits, risks, and best practices associated with peptide therapy.

Bibliography

For those interested in delving deeper into the science and application of peptides, the following bibliography provides a curated selection of seminal works and recent research findings that have shaped our understanding of peptide therapy. These resources offer a comprehensive look at the biochemical mechanisms, therapeutic potentials, and practical considerations of peptides in health and wellness. **"Peptides: Biology and Chemistry"** by James P. Tam and Peter T. P. Kaumaya offers an in-depth exploration of peptide synthesis, structure, and function, providing a solid foundation for understanding the role of peptides in biological systems. **"The Peptide Guide"** by Ryan D. Smith is an essential resource for anyone considering peptide therapy, covering everything from peptide selection and dosage to administration techniques and safety protocols. **"Clinical Applications of Peptides and Proteins in Aging and Neurodegenerative Diseases"** edited by Ahmet Alver and Tugba Gurpinar Tatar highlights the therapeutic applications of peptides in combating age-related conditions and neurodegenerative disorders, offering insights into the latest research and clinical studies. **"Peptide Applications in Biomedicine, Biotechnology and Bioengineering"** edited by Sotirios Koutsopoulos provides a comprehensive overview of the current and potential applications of peptides in various fields of medicine and biotechnology, emphasizing their role in innovative therapeutic and diagnostic approaches. For those seeking to understand the regulatory and ethical considerations surrounding peptide therapy, **"Regulatory Peptides and Cognate Receptors"** by Dianne L. Murphy and Mark A. Giembycz offers a detailed examination of the legal, ethical, and safety aspects of peptide use, including discussions on peptide sourcing, quality control, and regulatory compliance. Each of these texts contributes to a nuanced understanding of peptides, their therapeutic potential, and the practicalities of their use, making them invaluable resources for anyone looking to integrate peptides into their health and wellness regimen.

Analytical Index

For readers delving into the world of peptides with the objective of enhancing their health and wellness, the **Analytical Index** serves as a navigational tool, designed to streamline the search for specific topics and concepts covered in this comprehensive guide. This index meticulously categorizes the vast array of information presented, from the foundational science underpinning peptides and their myriad applications in anti-aging, muscle recovery, and overall wellness, to practical advice on their use, safety considerations, and legalities.

Key entries include detailed discussions on **peptide types** such as **GHRP**, **CJC-1295**, and **BPC-157**, each accompanied by insights into their mechanisms of action and therapeutic benefits. The index further breaks down complex topics like **cellular senescence**, **tissue regeneration**, and **lipolysis**, making it easier for readers to locate information on how peptides influence these biological processes.

Moreover, the index provides direct access to sections addressing **practical considerations** such as dosage guidelines, administration methods, and cycle planning, ensuring readers can easily navigate to the information necessary for informed, safe peptide use. It also highlights discussions on **side effects**, **contraindications**, and **health conditions** that warrant caution, offering a quick reference to safety information crucial for anyone considering peptide therapy.

Additionally, entries on **dietary** and **lifestyle integration** point readers towards strategies for maximizing peptide efficacy through nutrition, exercise, and rest, reflecting the holistic approach advocated throughout the book.

By facilitating swift access to these topics and more, the **Analytical Index** empowers readers to efficiently explore the content most relevant to their health goals and concerns, enhancing their ability to leverage the science of peptides for personal wellness.